T0220200

THE DECISION MAKER'S HANDBOOK TO DATA SCIENCE

A GUIDE FOR NON-TECHNICAL EXECUTIVES, MANAGERS, AND FOUNDERS

SECOND EDITION

Stylianos Kampakis

Apress®

The Decision Maker's Handbook to Data Science: A Guide for Non-Technical Executives, Managers, and Founders

Stylianos Kampakis
London, UK

ISBN-13 (pbk): 978-1-4842-5493-6 ISBN-13 (electronic): 978-1-4842-5494-3
https://doi.org/10.1007/978-1-4842-5494-3

Managing Director, Apress Media LLC: Welmoed Spahr
Acquisitions Editor: Shiva Ramachandran
Development Editor: Rita Fernando
Coordinating Editor: Rita Fernando

Cover designed by eStudioCalamar

Distributed to the book trade worldwide by Springer Science+Business Media New York, 233 Spring Street, 6th Floor, New York, NY 10013. Phone 1-800-SPRINGER, fax (201) 348-4505, e-mail orders-ny@springer-sbm.com, or visit www.springeronline.com. Apress Media, LLC is a California LLC and the sole member (owner) is Springer Science + Business Media Finance Inc (SSBM Finance Inc). SSBM Finance Inc is a **Delaware** corporation.

For information on translations, please e-mail rights@apress.com, or visit http://www.apress.com/rights-permissions.

Apress titles may be purchased in bulk for academic, corporate, or promotional use. eBook versions and licenses are also available for most titles. For more information, reference our Print and eBook Bulk Sales web page at http://www.apress.com/bulk-sales.

Any source code or other supplementary material referenced by the author in this book is available to readers on GitHub via the book's product page, located at www.apress.com/978-1-4842-5493-6. For more detailed information, please visit http://www.apress.com/source-code.

Printed on acid-free paper

Contents

About the Author

Dr. Stylianos (Stelios) Kampakis is a data scientist who is living and working in London, UK. He holds a PhD in Computer Science from the University College London as well as an MSc in Informatics from the University of Edinburgh. He also holds degrees in Statistics, Cognitive Psychology, Economics, and Intelligent Systems. He is a member of the Royal Statistical Society and an honorary research fellow in the UCL Centre for Blockchain Technologies.[1] He has many years of academic and industrial experience in all fields of data science like statistical modeling, machine learning, classic AI, optimization, and more.

Throughout his career, Stylianos has been involved in a wide range of projects: from using deep learning to analyze data from mobile sensors and radar devices, to recommender systems, to natural language processing for social media data, to predicting sports outcomes. He has also done work in the areas of econometrics, Bayesian modeling, forecasting, and research design. He also has many years of experience in consulting for startups and scale-ups, having successfully worked with companies of all stages, some of which have raised millions of dollars in funding. He is still providing services in data science and blockchain as a partner in Electi Consulting.

In the academic domain, he is one of the foremost experts in the area of sports analytics, having done his PhD in the use of machine learning for predicting football injuries. He has also published papers in the areas of neural networks, computational neuroscience, and cognitive science. Finally, he is also involved in blockchain research and more specifically in the areas of tokenomics, supply chains, and securitization of assets.

Stylianos is also very active in the area of data science education. He is the founder of The Tesseract Academy,[2] a company whose mission is to help decision makers understand deep technical topics such as machine learning and blockchain. He is also teaching "Social Media Analytics" and "Quantitative

---

[1]http://blockchain.cs.ucl.ac.uk/
[2]http://tesseract.academy

Methods and Statistics with R" in the Cyprus International Institute of Management[3] and runs his own data science school in London called Datalyst.[4]

Finally, he often writes about data science, machine learning, blockchain, and other topics at his personal blog: The Data Scientist.[5]

In his spare time, Stylianos enjoys (among other things) composing music, traveling, playing sports (especially basketball and training in martial arts), and meditating.

---

[3]www.ciim.ac.cy/
[4]www.dataly.st/
[5]http://thedatascientist.com/

Introduction

What is *data science*? What is *artificial intelligence*? What is the difference between *artificial intelligence and machine learning*? What is the best algorithm to use for X? How many people should I hire for my data science team? Do I need a recommender system? Is a *deep neural network* a good idea for this use case?

Having dedicated my career to understanding data, and modeling uncertainty, these questions (and many similar ones) have popped up very often in conversations I am having with CEOs, startup founders, and product managers. There are always three common elements:

1. The people involved have a non-technical background.

2. Their business collects data, or is in a position to collect data.

3. They want to use data science but they don't know where to start.

Data science (and all the fields it encompasses such as AI and machine learning) can transform our world on every level: business, political, and individual. In more than one way, this is already happening. Online retailers know what you are going to like, through the use of recommendation engines. Your photographs get automatically tagged through the use of computer vision. Autonomous vehicles can drive us around with no driver in the seat.

However, this is still only a fraction of the things that are possible with data science. The benefits of this powerful technology will never be reaped, unless the entrepreneurs and the decision makers fully understand how to use it.

Data science is unique in the space of technology in two ways. First in contrast to software development, it is intangible. You can't see a flashy front end, but only the results of a model. Secondly, it is *science*, which means that, in contrast to engineering, it is difficult to set out a perfectly laid plan in advance. Uncertainty is an integral part of data science, and this can make estimations and decisions more difficult. These two factors make the understanding of data science more challenging.

The cloud of buzzwords currently dominating technology does not really help at all with that. I have seen buzzwords such as "big data," "analytics," "prediction," "forecasting," and many more used without any real understanding of the context surrounding them. This is partly due to aggressive sales tactics that end up confusing rather than illuminating. The result is that many entrepreneurs end up feeling more insecure, since they can't understand what they need and how much is a fair price to pay. This is why I realized that it is upon us, the data scientists, to take up the role of educators.

This book is meant to be the ultimate short handbook for a decision maker who wants to use data science but is not sure where to start. All case studies outlined are always described with the decision maker in mind. The problem in business is not how to choose the right model from a scientific viewpoint but how to deliver *value*. The data scientist has to make decisions based on trade-offs such as the cost of development (which can include time and hiring), the interplay with business decisions, and the cost of data. The book explains how the decision maker can better understand these dilemmas and help the data scientist make the most beneficial choices for the business.

I hope that after reading this book, the world of data science will no longer be a dangerous landscape dominated by buzzwords and incomprehensible algorithms, but rather a place of wonder, a place where the future lies. I do hope that you will enjoy reading it as much as I enjoyed writing it.

Demystifying Data Science and All the Other Buzzwords

In the business world, data has become a big thing. You hear all sorts of buzzwords being thrown around left, right, and center. Things like *big data*, *artificial intelligence*, *machine learning*, *data mining*, *deep learning*, and so on. This can get confusing, leaving you paralyzed as to what is the best technology to use and under what circumstances. In this chapter, we are going to demystify all these buzzwords, by taking a short stroll through the history of data science. You will understand how the history of data analysis gave birth to different schools of thought and disciplines, which have now all come together under the umbrella of the term **data science**.

© Stylianos Kampakis 2020
S. Kampakis, *The Decision Maker's Handbook to Data Science*,
https://doi.org/10.1007/978-1-4842-5494-3_1

What Is Data Science?

In 2017, we generated more data than we did over the previous 5000 years of our history.[1] That's a lot of data. And it's not surprising. Every device we own generates data, and all our interactions with said devices generate even more data.

So, you take a picture with your smartphone? You've generated data. You read the news on your tablet? You're generating data. You listen to a podcast on your laptop? You're generating data. You go on Facebook to update your status? You've generated data.

You get the point. There's a good chance that the only thing we do that doesn't generate data is breathe, but even that is debatable considering all the wearable devices that are available and which can track everything from heart rate to calories burned.[2]

What happens to all this data? It must have some use or things wouldn't be set up so that we generate it in the first place.

At the moment, some of it does and some of it doesn't. In fact, some studies have discovered that less than 0.5% of data is being analyzed and turned into actionable insights.[3] But more on that later.

The important thing is that all data *can* be used. We just need to figure out better ways of doing so.

And this is where data science comes in. There are many definitions of data science. One such definition is

> Data science is a "concept to unify statistics, data analysis, machine learning and their related methods" in order to "understand and analyze actual phenomena" with data.[4]

Okay, so those are a lot of big and fancy words, which can get confusing. Let's boil it down to the simplest definition:

> Data science is about using data to do useful stuff.

---

[1] David Sønstebø, "IOTA Data Marketplace," *IOTA*, November 28, 2017, https://blog.iota.org/iota-data-marketplace-cb6be463ac7f

[2] www.wired.co.uk/article/hospital-prescribing-tech

[3] www.technologyreview.com/s/514346/the-data-made-me-do-it/

[4] Chikio Hayashi, "What is Data Science? Fundamental Concepts and a Heuristic Example," in *Data Science, Classification, and Related Methods*, eds. Chikio Hayashi, Keiji Yajima, Hans-Hermann Bock, Noboru Ohsumi, Yutaka Tanaka, and Yasumasa Baba (Tokyo, Japan: Springer-Verlag, 1998), 40-51.

Short and to the point. That's exactly what data science is all about. The methods we use are important, of course, but the essence of this discipline is that it allows us to take data and transform it so we can use it to do useful things.

For example, let's say someone visits a doctor because they are short of breath. They are experiencing heartburn and they have chest pains. The doctor will run the basic tests, including measuring blood pressure, but nothing seems out of the ordinary.

Since the patient is overweight, the doctor immediately assumes the symptoms are caused by the patient's size and recommends a healthier diet and exercise.

Three months later, the same patient is brought into the emergency room and ends up dying on the table because of a heart defect.

This might sound like an episode on your favorite medical show, that is, a work of fiction, but it happens much more often than you might think. In fact, in the United States, 5% of patients are misdiagnosed, while misdiagnosis cost the United Kingdom over £197 million in the 2014/2015 fiscal year.[5]

However, this situation can be avoided thanks to data science. The analysis of similar cases reveals that the symptoms our patient exhibited aren't just caused by obesity but also by some cardiovascular conditions.

Having access to systems that can analyze data and compare it with new data inputs could have helped the doctor identify the problem sooner without relying solely on their own personal experience and knowledge.

So, data science can be used to save lives, among many other applications, which is pretty useful.

Data Science Is Multidisciplinary

Data science involves multiple disciplines, which is why finding someone with the necessary skills to be a data scientist can be difficult.

Thus, data science involves everything from statistics and pattern recognition to business analysis and communication. It requires creative thinking as much as it requires analytical thinking.[6]

---

[5] Lena Sun, "Most Americans Will Get a Wrong or Late Diagnosis At Least Once In Their Lives," *Washington Post*, September 22, 2015, www.washingtonpost.com/news/to-your-health/wp/2015/09/22/most-americans-who-go-to-the-doctor-will-get-a-wrong-or-late-diagnosis-at-least-once-in-their-lives-study-says/; "The Top Misdiagnosed Conditions In NHS Hospitals In 2014/15", *Graysons*, www.graysons.co.uk/advice/the-top-misdiagnosed-conditions-in-nhs-hospitals/
[6] Take a look at this infographic by Brendan Tierney: https://oralytics.com/2012/06/13/data-science-is-multidisciplinary/

So, data science involves discovering which data is useful as well as effective ways of managing it. It also requires determining how the data should be processed and what types of insights can be garnered from the massive amounts of data available.

Data science requires knowledge of programming and computing, but also visualization so that the insights can be presented in a way that everyone can understand.

Furthermore, business acumen is also a necessity because while data science can be applied to any field of business, it is critical to know what types of answers the business needs and how to present said insights so leadership can understand them.

Core Fields of Data Science

Data science has three core fields, namely, artificial intelligence, machine learning, and statistics.

Artificial intelligence is all about replicating human brain function in a machine. The primary functions that AI should perform are logical reasoning, self-correction, and learning. While it has a wide range of applications, it is also a highly complicated technology because to make machines smart, a lot of data and computing power is required.

Machine learning refers to a computer's ability to learn and improve beyond the scope of its programming. Thus, it relies on creating algorithms that are capable of learning from the data they are given. They are also designed to garner insights and then make forecasts regarding data they haven't previously analyzed.

There are three approaches to machine learning, namely, supervised, unsupervised, and reinforcement learning, plus some subfields (such as semi-supervised learning). Here, we will be talking only about supervised and unsupervised learning, since this is what is mainly used in business.

Let's say you want to sort all your photographs based on content. In supervised learning, you would provide the computer with labeled examples. So, you'd give it a picture of a dog and label it animal. Then you'd feed it a picture of a person and label it human. The machine will then sort all the remaining pictures.

In unsupervised learning, you'd just give the machine all the photos and let it figure out the different characteristics and organize your photos.

In reinforcement learning, the machine learns based on errors and rewards. Thus, the machine analyzes its actions and their results. A good example of reinforcement learning is a rat in a maze that needs to navigate its way to a piece of cheese. The learning process that helps the rat achieve this can be implemented in a machine through reinforcement learning. This is one of the most esoteric types of machine learning.

Statistics is an essential tool in the arsenal of any data scientist because it helps develop and study methods to collect, analyze, interpret, and present data. The numerous methodologies it uses enable data scientists to

- Design experiments and interpret results to improve product decision-making
- Build signal-predicting models
- Transform data into insights
- Understand engagement, conversions, retention, leads, and more
- Make intelligent estimations
- Use data to tell the story

Let's take a closer look at all three.

Artificial Intelligence: A Little History

In 1954, the field of AI research came into being at a workshop at Dartmouth College. There, the attendees discussed topics that would influence the field for years to come.

As we've already explained, the goal of artificial intelligence is to create a "thinking machine," that is, one that emulates human brain function. To do this, of course, one needs to understand the human mind, which is why AI is closely related to the field of cognitive science.

Cognitive science involves studying the human mind and its processes, including intelligence and behavior. Memory, language, perception, attention, emotion, and reasoning are all studied, and to gain greater understanding of these faculties, scientists borrow from other fields, including

- Linguistics
- Psychology
- Philosophy
- Neuroscience
- Anthropology
- Artificial intelligence

Artificial intelligence played a key part in the development of cognitive science, and there was a large interplay between cognitive psychology and AI. The understanding of human cognition helped us improve our understanding of how to transfer this inside machines. Vice versa, the computational theory of the mind[7] was one of the dominant paradigms in cognitive science. According to this theory, the mind works like a computer, with processes and limited memory. While this is now considered outdated, it drove research for decades.

The AI Dream

It all started with a grand vision. Marvin Minsky, the head of the artificial laboratory at MIT who was considered the father of AI, stated that "artificial intelligence is the science of making machines do things that would require intelligence if done by men."[8] In other words, the dream was to create an intelligent machine.

So, how would one go about doing it? First, we start with intuition, that is, the human's distinct ability for logical reasoning.

Logical reasoning is, essentially, the capacity to reason based on various premises to reach a conclusion that has logical certainty. For example:

If all men are mortal, and Socrates is a man, then Socrates is mortal.

Or:

If it's raining outside and I don't have an umbrella, I will get wet if I go out.

To translate this logical deduction ability to a machine, a rule-based or symbolic approach is used. This involves humans constructing a system of rules with which the computer is programmed. Using these rules, reasoning algorithms are capable of deriving logical conclusions.

A good example is MYCIN,[9] which was an early successful working system based on reasoning algorithms. It was used to diagnose infections and determine the type of bacteria that was causing the problem. It was never used in a clinical setting but is an excellent example of an expert system and a predecessor to machine learning.

---

[7] https://plato.stanford.edu/entries/computational-mind/
[8] Blay Whitby, *Reflections on Artificial Intelligence* (Exeter, UK: Intellect Books, 1996).
[9] A good old reference for MYCIN by John McCarthy can be found here: www-formal. stanford.edu/jmc/someneed/someneed.html

The system was developed in the 1970s at Stanford University and had approximately 600 rules.[10] Users were required to provide answers to various questions and the program would then provide a list of potential bacteria that could be causing the problem, sorted from high to low probability. It would also provide its confidence in the probability of each diagnosis as well as how it came to the conclusion. Finally, it would provide the recommended course of treatment.

It had a 69% accuracy rate, and it was claimed that the program was more effective than junior doctors and on the same level as some experts.[11]

The program was created by interviewing a large number of experts who provided their expertise and experience. It used rules of the IF (condition) THEN (conclusion) form. For example, IF (sneezing and coughing or headache) THEN (flu).

One limitation the program had was computing power. It took approximately 30 minutes to go through the system, which was too much wasted time in a real-world clinical setting.

Another issue was also raised, namely, that of ethics and legal issues. Thus, the question arose of who would be held responsible of the program made the wrong diagnosis or recommended the wrong treatment.

Though it was never used, MYCIN still had a very important role in bringing us to where we are today as it was one of the early successes of AI, proving what is possible and strengthening.

Automated Planning

Planning is a vital component of rational behavior. Automated planning and scheduling is an area of artificial intelligence that involves the creation of a system that is capable of selecting and organizing actions to achieve a certain outcome.

An example is the Missionaries and Cannibals problem,[12] which is a classic AI puzzle. It is defined as follows:

> On one bank of a river are three missionaries and three cannibals. They all wish to cross to the other side of the river. There is one boat available that can hold up to two people. However, if the cannibals ever outnumber the missionaries on either of the river's banks, the missionaries will get eaten.

[10] Bruce G Buchanan and Edward H Shortliffe, *Rule-Based Expert Systems* (Reading, MA: Addison-Wesley, 1985).

[11] Victor L. Yu, "Antimicrobial Selection By A Computer," *JAMA* 242, no. 12 (1979): 1279, https://jamanetwork.com/journals/jama/article-abstract/366606

[12] You can play a version of the game here: www.novelgames.com/en/missionaries/

> How can the boat be used to safely carry all the missionaries and cannibals across the river?

It's a little gruesome and might also seem trivial, but a similar approach can be used for schedule planning.

Other similar problems in that are the towers of Hanoi[13] and the travelling salesman problem. The travelling salesman problem is a legendary benchmark in optimization, where the objective is to find a path for a salesman to go across all the cities in a country (or some other geographical regions). This path should never go through the same city twice, and at the same time it should as short as possible. It is easy to see how this relates to vehicle routing in real life. In Figure 1-1, you can see an example of the travelling salesman for all major cities in Germany.

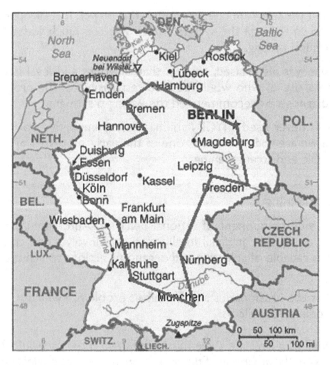

Figure 1-1. Example of the travelling salesman problem

Essentially, what the computer does is analyze each possibility, discarding the one that doesn't fulfill the parameters while presenting all the options that do.

---

[13] www.mathsisfun.com/games/towerofhanoi.html

The AI Winters

Artificial intelligence would likely be much further along now were it not for the "winters" it experienced. The term AI winter refers to periods when interest in artificial intelligence was diminished, and, as a result, funding was limited.

In 1967, Marvin Minsky predicted that "within a generation the problem of creating 'artificial intelligence' will be substantially solved." However, by 1982, he admitted that "the AI problem is one of the hardest science has ever undertaken."[14]

The first major AI winter started in 1974 and ended in 1980, while the second one started in 1987 and ended in 1993. There were other smaller issues, such as

- 1966—The failure of machine translation
- 1970—The abandonment of connectionism
- 1971-1975—DARPA's frustration with the Speech Understanding Research program at Carnegie Mellon University
- 1973—The large decrease in AI research in the United Kingdom in response to the Lighthill report
- 1973-1974—DARPA's cutbacks to academic AI research in general
- 1987—The collapse of the Lisp machine market
- 1988—The cancellation of new spending on AI by the Strategic Computing Initiative
- 1993—Expert systems slowly reaching the bottom
- 1990s—The quiet disappearance of the fifth-generation computer project's original goals

The first winter was caused, in large part, by three major elements. First, it was the Lighthill report, in which professor Sir James Lighthill concluded that other sciences could do everything being done in AI, implying that many of the most successful AI algorithms would be incapable of solving real-world problems.[15]

---

[14]Frederick E. Allen, "The Myth Of Artificial Intelligence | AMERICAN HERITAGE," *Americanheritage.Com*, last modified 2001, www.americanheritage.com/content/myth-artificial-intelligence

[15]Lighthill, Professor Sir James (1973). "Artificial Intelligence: A General Survey." *Artificial Intelligence: a paper symposium*. Science Research Council.

Though contested, the report still resulted in the complete shutdown of AI research in the United Kingdom, with only a few universities continuing the research. The result was funding cuts for this research all across Europe.[16]

In 1969, the Mansfield Amendment was passed in the United States, requiring DARPA to fund research that had a clear mission rather than basic projects with no clear direction. Essentially, researchers had to prove their projects would quickly produce technology that would be useful to the military.

In concert with Lighthill's report, which was used as proof that AI research was unlikely to provide anything useful in the foreseeable future, DARPA stopped funding this type of research. By 1974, funding for AI was virtually impossible to find.[17]

Another issue was the Speech Understanding Research program at Carnegie Mellon. DARPA wanted a system that could respond to spoken commands from a pilot, and while the team had developed a system that understood spoken English, the words had to be spoken in a certain order.

DARPA felt they had been misled and cancelled the grant in 1974.[18] Interestingly enough, by 2001 the speech recognition market reached $4 billion and used the technology the team from Carnegie Mellon developed.[19]

In 1980, the commercial success of expert systems rekindled interest in the field of AI. By 1985, over a billion dollars was being spent on AI and an industry developed to support the field. Specialized computers dubbed Lisp machines were built to better process the programming language that was preferred for AI.

Unfortunately, the Lisp machine market collapsed because companies like Sun Microsystems developed more efficient alternatives. These general workstations had much better performance, which Lisp machines were incapable of matching. This led to the second AI winter.

What We Learned from AI Research

Despite all the issues, research into artificial intelligence discovered many beneficial things. For example, if it weren't for AI research, we would not have algorithms that work with logical rules.

---

[16] Daniel Crevier, *The Tumultuous History Of The Search For Artificial Intelligence* (New York, NY: Basic Books, 1993); Russell & Norvig 2003; Jim Howe, "School Of Informatics: History Of Artificial Intelligence At Edinburgh," *Inf.Ed.Ac.Uk*, last modified 2007, www.inf.ed.ac.uk/about/AIhistory.html

[17] National Research Council Staff, *Funding a Revolution* (Washington: National Academies Press, 1999).

[18] Crevier 1993; McCorduck 2004; National Research Council 1999.

[19] National Research Council 1999.

An early example of such an algorithm was the Prolog language. One of the first logic-based programming languages, it's still one of the most popular even today. It has been used to prove theorems, in expert systems, in type inference, and for automated planning.

And that's another area where AI has helped us. Automated planning and scheduling, which is used to drive things like automated robots in manufacturing, would not exist otherwise. Neither would expert systems or knowledge representation.

Of course, all these things have evolved significantly since then, and we are getting closer and closer to creating true AI, but we still have a long way to go before we have a truly intelligent machine.

Another interesting aspect is that cognitive science borrows from AI as much as the reverse is true. The desire to develop a smart machine that can think like a human required a better understanding of the human mind. This led to research into memory, perception, and much more.

The Next Step: Enter Machine Learning

As a term, machine learning was coined by Arthur Samuel, who was a pioneer in computer gaming and AI, in 1959 while working at IBM.[20]

He defined it as follows:

Machine learning is about giving computers the ability to learn without being explicitly programmed.

A rift caused by the focus on logical, knowledge-based approaches developed between machine learning and AI. By the 1980s, expert systems dominated AI, and statistics was no longer of interest.

Machine learning was then reorganized as a separate field and began to gain serious traction in the 1990s. Instead of focusing on creating artificial intelligence, machine learning shifted its approach and concentrated on solving more practical problems. Thus, instead of employing the symbolic methodologies so prevalent in AI, it began to employ models taken from statistics and probability theory.

While machine learning as a concept is a subfield of artificial intelligence, its approach was practically opposite to research that had been done so far.[21]

---

[20] Arthur L. Samuel, "Some Studies In Machine Learning Using the Game of Checkers", *IBM Journal of Research and Development* 44, no. 12 (1959): 206-226, http://citeseerx. ist.psu.edu/viewdoc/download?doi=10.1.1.368.2254&rep=rep1&type=pdf
[21] http://thedatascientist.com/machine-learning-vs-ai/

Thus, "classic AI" or "good-old-fashioned AI" took a ruled-based, top-down approach. Essentially, it meant giving computers rules that they had to follow to reach the desired outcome. The problem with this approach is that it requires a lot of human input.

The designer of the system must be able to predict every possibility and create a set of rules for it. Furthermore, uncertainty is not easily handled because the machine thinks in an IF/THEN pattern.

Unfortunately, the world isn't always so clear-cut. In the real world, IF (EVENT A) takes place, THEN (ACTION B) isn't always the solution. For example, IF (SUNNY) doesn't always result in THEN (WEAR LIGHT CLOTHING). It could be winter out and you could need a heavy coat. Furthermore, the system only knows to provide this answer because it's been told what to do.

Machine learning, on the other hand, takes a bottom-up, data-driven approach. So, instead of programming a computer with rules it must follow, machine learning uses data to teach the computer what to do.

There is very little human input involved, other than providing the data. This allows the machine to handle uncertainty more naturally, making it a more effective approach since the world is fundamentally more probabilistic than certain.

With our rain analogy, machine learning would work a little like this. The system is fed a ton of data and it analyzes what happens when there is a certain type of weather. So, when it's sunny, it sees people going out in both light and heavy clothing. It learns even more when it makes the connection that temperature also matters. Essentially, it recognizes patterns: sun + cold temperature = warmer clothing. The machine does this on its own by analyzing and learning from data it has been given.

The problem with machine learning for a while was that it required a lot of data to work effectively. Thankfully, with technological progress like the development of the Internet and cheap storage, the data-driven approach has become more popular. Not only is it more effective, but it's also more feasible from an economic and technological viewpoint too.

The Problem with Machine Learning

Of course, machine learning has its drawbacks, just like anything else. First of all, machine learning can't handle logic rules directly, and how we can integrate machine learning with reasoning is an open question.

Let's take gravity as an example. Machine learning can come to the conclusion that there is a 0.99 probability of an object falling if you let it go. This, however, is based on past data and it cannot make the logical leap that "any" object will fall if you let it go because of gravity. It can give probabilities based on past events, but it cannot come to the conclusion: IF (GRAVITY) THEN (OBJECT FALL). This logical leap is something that even babies can do, demonstrating that while machine learning can be very powerful, it is far from intelligent the way the inventors of AI envisioned it.

Another issue is that machine learning isn't all that transparent. Most algorithms are a "black box," with input going in on one side and the results being spat out on the other side. However, we don't see what's happening inside the box. Think of Google's famous search algorithm. We know we enter a search string and we get the "most" relevant results, but we don't see how that happens.

Interpretable machine learning is now a very active field of investigation.[22]

Deep Learning

Deep learning is the field of research that studies *deep neural networks*, which is a neural network with multiple layers. A neural network will have one or two layers, whereas deep neural networks have many more (Figure 1-2).

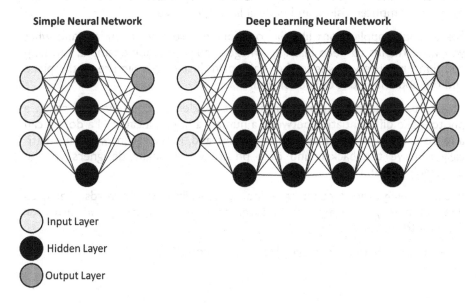

Figure 1-2. Example of a regular neural network vs. a deep neural network

---

[22] http://thedatascientist.com/interpretable-machine-learning

Geoffrey Hinton, a British cognitive psychologist and computer scientist who is regarded as the "Godfather of Deep Learning," discovered it was impossible to get his research into neural networks published, so he had to devise a new term that was "cooler."[23]

Neural networks consist of neurons and the pathways that connect them in the human brain. Since the brain is the most powerful computing machine known to man, it makes sense to attempt to emulate its architecture, which led to the creation of artificial neural networks.

Neural networks have traditionally consisted of three layers: an input layer, a middle layer, and an output layer. The middle layer is what provided the neural networks with the power to perform non-linear computations and turned them into powerful learning machines.

The idea behind deep learning had been around for decades.[24] However, it was believed that adding more layers was not really useful. Furthermore, it was difficult to train these networks.

Deep learning takes advantage of artificial neural networks to enable machines to learn faster, more accurately, and at scale. Thus, it is used to help develop solutions to many issues that were previously unsolvable, such as speech recognition, computer vision, and natural language processing.

For example, applications like Siri or Cortana are able to understand what you want thanks to deep learning. Image recognition, for example, can help you find furniture you see in the real world. Just take a picture of the piece and the app will find it online for you.

Thanks to deep learning, machines can also identify different subjects in a photo—like differentiating between a human, a horse, a cat, and a dog. It can also be used to automatically color photos and, more recently, to automatically remove the background of a photo, which many designers will be extremely grateful for.

It is also being used to generate text automatically. In other words, thanks to deep learning, your computer can listen to what you're saying and turn your words into text.

Basically, deep learning can do a lot of really cool stuff!

---

[23] Chris Sorensen, "How U Of T's 'Godfather' of Deep Learning Is Reimagining AI," *University of Toronto News*, November 2, 2017, www.utoronto.ca/news/how-u-t-s-godfather-deep-learning-reimagining-ai; Adrian Lee, "Geoffrey Hinton, The 'Godfather' of Deep Learning, on AlphaGo," *Macleans*, March 18, 2016, www.macleans.ca/society/science/the-meaning-of-alphago-the-ai-program-that-beat-a-go-champ/
[24] http://thedatascientist.com/what-deep-learning-is-and-isnt/

But how does it do that and why it works so well? The multiple layers of a deep neural network allow it to extract *hierarchical features*. Our visual system is organized hierarchically. Our brain picks up simple blobs and edges before it picks up shapes (such as squares and circles) and then moves on to facial features or other more advanced concepts. Deep neural networks essentially mimic this functionality. If we take a look at the kind of processing that the middle layers of deep neural network are doing, then we can see that they extract a hierarchy of features. This is why deep neural networks have been so successful in computer vision and language applications. These are domains where by nature are dominated by hierarchical features. ConvNet Playground[25] has a model explorer that allows you to see how the different layers of Inception (a famous deep neural network) create more and more complicated representations of an image as the network gets deeper.

However, deep neural networks are not a cure-all.[26] If the domain is not characterized by multiple interacting variables, or by hierarchies of features, or there is not enough data, then deep neural networks are probably an overkill. So, if, for example, you have a dataset of 1000 rows and 10 variables, you are much better off using a simpler model (e.g., linear regression), since this model will train faster, and it is less likely to *overfit*.

Overfitting[27] is the bane of all machine learning algorithms and refers to a situation where the performance in the real world is much worse than during training. More complicated models can suffer from this problem. This is why deep neural networks require excessive tuning and testing to make sure they work well.

Statistics

The other core field of data science, statistics, is a branch of mathematics that started around the 18th century and was the first attempt to systematically analyze data. It produced the cornerstones of all modern data analysis, including summary statistics, regression and classification, significant testing, forecasting, and more.

---

[25] https://convnetplayground.fastforwardlabs.com/#/models
[26] James Somers, "Progress in AI Seems Like It's Accelerating, But Here's Why It Could Be Plateauing," *MIT Technology Review*, September 29, 2017, www.technologyreview.com/s/608911/is-ai-riding-a-one-trick-pony/
[27] www.quora.com/What-is-an-intuitive-explanation-of-over-fitting-particularly-with-a-small-sample-set-What-are-you-essentially-doing-by-over-fitting-How-does-the-over-promise-of-a-high-R%C2%B2-low-standard-error-occur

What Makes Statistics Unique?

Statistics generally deals with smaller samples, in large part due to historical reasons. The Internet makes it possible to collect a lot of data nowadays, but in the "olden" days, data usually had to be collected in person, which is why sample sizes were so small.

The results were then extrapolated to the entire population with a margin of error. Also, consider that it was developed in the 17th century, which meant everything had to be calculated by hand.[28] Solving a model would take days or weeks.

Statistics is also very strict on model assumptions and development, because it is rooted in math. It's almost like classic AI in this sense. This gap did narrow after parts of machine learning were founded on Bayesian statistics.

However, the mentality still differs. In machine learning, it's all about getting it done and asking questions later. In statistics, though, assumptions must be validated and verified before anything else can be done.

While statistics offers a higher degree of transparency and interpretability, machine learning has led to incredible performance. This amazing performance has allowed us to perform tasks previously thought impossible!

Of course, one has to remember that machine learning is the product of the computer era, which means much faster calculations. Instead of it taking days of work for a single model, it takes minutes—if that—and multiple approaches can be tried in a single day.

The Battle: Statistics vs. Machine Learning

Some statisticians are incredibly opposed to machine learning.[29] Religiously opposed. Sometimes, it's actually a bit worrisome. Their problem is the lack of transparency in machine learning as well as a lack of theory.

However, in reality, statistics and machine learning are two sides of the same coin. Furthermore, they are tools in a data scientist's toolbox and can be applied as the situation requires. So, if you are looking for interpretability, then statistics is your best bet, while predictive power will be better served by machine learning.

---

[28] Walter F. Willcox, "The Founder of Statistics," *Revue de l'Institut International de Statistique/Review of the International Statistical Institute* 5, no. 4 (1938): 321, www.jstor.org/stable/1400906

[29] http://thedatascientist.com/statistics-vs-machine-learning-two-worlds/

Subfields of Data Science

In parallel to the development of the aforementioned core fields of data science, other fields were also being developed. These were all related in some way to data analysis, artificial intelligence, and so on. Such fields include cybernetics developed by Norbert Wiener in 1947, artificial neural networks thanks to McCulloch and Pitts in 1943, computational intelligence developed in 1990, machine learning developed in 1959, and data mining and knowledge discovery in databases thanks to Gregory Piatetsky-Shapiro in 1990.[30]

The problem with all these terms is that not everyone perceives them in the same way necessarily or the relationship between them. In some cases, there are conceptual differences between them, but sometimes it's a matter of who the person talking about them is and what their research background is. There's also the matter of marketing, which can also give a different spin to a particular subfield.

For example, some people consider AI as the core field with natural language processing, expert systems, neural networks, fuzzy logic, and robotics as satellite fields. Others consider that AI is part of computer science, while curated knowledge, machine learning, and reverse engineering the brain are essential to AI. Yet others believe machine learning, deep learning, and representation learning to be contained within AI.

In other words, people have different interpretations of what all these technical terms mean and how they relate to each other, which can make it very confusing. To further prove my point, the following is a definition for computational intelligence at the 11th UK Workshop on Computational Intelligence:

> *Computational Intelligence (CI) is an offshoot of artificial intelligence in which the emphasis is placed on heuristic algorithms such as **fuzzy systems, neural networks** and **evolutionary computation**. It is usually contrasted with 'traditional', 'symbolic' or 'good old fashioned artificial intelligence (GOFAI)'. The IEEE Computational Intelligence Society uses the tag-line '**Mimicking Nature for Problem Solving**' to describe Computational Intelligence, although mimicking nature is not a necessary element.*

---

[30] Flo Conway and Jim Siegelman, Dark Hero of the Information Age (New York: Basic Books, 2006); Warren S. McCulloch and Walter Pitts, "A Logical Calculus Of The Ideas Immanent In Nervous Activity," *The Bulletin of Mathematical Biophysics* 5, no. 4 (1943): 115-133, www.cse.chalmers.se/~coquand/AUTOMATA/mcp.pdf; Russell & Norvig 2003, pp. 25–26; Samuel 1959; Gregory Piatetsky-Shapiro, "Knowledge Discovery In Databases: Progress Report," *The Knowledge Engineering Review* 9, no. 01 (1994): 57.

*In addition to the three main pillars of CI (fuzzy systems, neural networks and evolutionary computation), Computational Intelligence also encompasses elements of **learning, adaptation, heuristic** and **meta-heuristic optimization**, as well as any hybrid methods which use a combination of one or more of these techniques. More recently, emerging areas such as artificial immune systems, **swarm intelligence, chaotic systems**, and others, have been added to the range of Computational Intelligence techniques. The term 'Soft Computing' is sometimes used almost interchangeably with Computational Intelligence.[31]*

Now, let's take a look at the definition for Knowledge Discovery in Databases:

*The term Knowledge Discovery in Databases, or KDD for short, refers to the broad process of finding knowledge in data, and emphasizes the "high-level" application of particular data mining methods. It is of interest to researchers in **machine learning, pattern recognition, databases, statistics, artificial intelligence, knowledge acquisition for expert systems, and data visualization**. The unifying goal of the KDD process is to extract knowledge from data in the context of large databases.[32]*

The bolded terms show the overlap between all the fields and subfields to the point where some of the core fields are considered subfields of the subfields. Yes, it's confusing and that's exactly the point.

All these fields are interdisciplinary in the sense that they borrow tools from each other, and though they might sometimes take a different approach, there is still so much overlap that it is incredibly problematic. It's even confusing for the people who study these fields.

Take *Pattern Recognition and Machine Learning* by Christopher M. Bishop,[33] which was one of the most important textbooks for machine learning. Everyone has read it, and yet, one has to wonder at the difference between pattern recognition and machine learning. However, once you get into it, you realize that it gets into statistical theory and, generally, just makes things much more complicated.

And that's not even close to the end of the matter. There are many other related popular buzzwords and fields, such as

- Natural language processing, which is a subfield of machine learning focused solely on language—generally it relies on text but can also include speech.

---

[31]"UKCI 2011—Introduction: What Is Computational Intelligence," http://ukci.cs.manchester.ac.uk/intro.html

[32]University DBD, "KDD Process/Overview," www2.cs.uregina.ca/~dbd/cs831/notes/kdd/1_kdd.html

[33]Christopher M. Bishop, *Pattern Recognition and Machine Learning* (New York, Springer).

- Recommender systems, which is a field dedicated to the study of recommendation engines like the ones used by Amazon and Netflix.

- Predictive analytics, which isn't a sub-discipline but a term that refers mainly to the development of predictive models, that is, algorithms that can predict the future. The latter can be developed using either machine learning or statistics.

In other words, it's all really, really confusing, which is where the term "data science" comes in.

Data Science: The Skills Needed

Data science makes it much easier to explain everything, unless one wants to spend half an hour explaining all the fields and subfields. Even then, most people aren't going to understand, especially if they're not in related fields.

It's much simpler to say one is a data scientist who does stuff with data. It's certainly a lot better than going into a lengthy litany about knowing machine learning, data mining, statistics, and so on.

However, even data science is not an infallible term and can cause a little confusion. Some might think that data engineering, databases, big data, and data analysts are pretty much the same thing. But they aren't. So, what are the differences? Well, let's take a look at what these terms mean.

- *Data Scientist*

   A data scientist is someone who works with data. However, this term has been expanded lately to refer to someone who analyzes data but who also has some form of advanced training in a related field, such as a PhD in machine learning or a master's in statistics.

- *Data Engineer*

   A data engineer is someone who designs, builds, and manages the infrastructure for the information of big data. This is the person who develops the architecture that helps analyze and process data. To make a cooking analogy, the data scientist is the chef, while the data engineer encompasses all the tool manufacturers (knives, pots, pans, stoves, etc.) and ingredient vendors who provide the chef with everything he or she needs to create a dish.

- *Big Data*

  Big data is a buzzword as it simply refers to massive data-sets that are so large and complicated; traditional data processing application software simply can't handle them. It relates to NoSQL databases, cloud computing, Hadoop, and so on.

Table 1-1 shows the important skills for the different categories of data professionals.

Table 1-1. Important skills for different categories of data professionals, ranked out of 3.

	Data analyst	Machine learning engineer	Data engineer	Data scientist
Programming tools	3/3	1/3	3/3	3/3
Data intuition	3/3	3/3	1/3	3/3
Data visualization	3/3	2/3	1/3	3/3
Statistics	2/3	2/3	0/3	3/3
Machine learning	0/3	3/3	0/3	3/3
Software engineering	0/3	2/3	3/3	2/3
Data wrangling	0/3	3/3	3/3	3/3
Mathematics	0/3	3/3	0/3	2/3
Databases	0/3	1/3	3/3	2/3

What Does a Data Scientist Need to Know?

We will delve much more into data scientists later on in this book, but here is a quick overview.

While data scientists can come from varied backgrounds, you will usually find three types.

First you have the statisticians, some of which don't like to be called data scientists.

Then you have people with computer science backgrounds who studied machine learning and have either master's degrees or PhDs in the field.

Lastly, you have people coming from other disciplines, like physics; you learned to use data science tools and migrated into the field.

THE QUESTION OF GENERAL ARTIFICIAL INTELLIGENCE

One of the hottest trending topics of our time is "general artificial intelligence." This term refers to the creation of artificial intelligence that can be just as smart as we are. People who believe that this is achievable in the foreseeable future also believe in the concept of a "singularity." A singularity in physics is defined as a moment when the physical models break down (e.g., a situation that exists in the center of a black hole). A technological singularity is a situation where everything we know about society and technology will break down, and something new and unpredictable will show up.

This is based on the following kind of reasoning. While we, humans, possess intelligence, we are limited in our cognitive capacity. We can't hold an infinite size of information in our brain. We can think fast, but not faster than a computer makes calculations. Also, we can get tired, and we need to sleep.

A general artificial intelligence will have our own ability to think, but none of our limitations. Hence, if a machine had the intelligence of a gifted mathematician, then, in theory, it could come up with a huge number of mathematical theorems within seconds. It could develop capabilities and make inventions that would take humans many lifetimes to achieve.

There are people, like Elon Musk or the late Stephen Hawking, who have described AI as a threat to humanity. On the other hand, other people are more optimistic about the potential of general AI to change life for the better.

Irrespective of which side is true, there are a few things that all experts agree on now:

1. We don't have a clear path to general AI. There is not a single method or technology that is guaranteed to achieve this result.

2. Based on the current technological progress, it is very likely that we will see general AI being achieved within our century.

3. While the results of general AI might not be as dramatic as some people expect, everyone agrees that there will be huge disruption in multiple areas: from health research to military.

We are definitely living in interesting times!

Data Management

Data management includes all the disciplines involved in working with data and treating it as the valuable resource that it is. A more formal definition is provided by DAMA International,[1] which is the professional organization for the profession of data management. Thus, their definition is as follows:

> Data Resource Management is the development and execution of architectures, policies, practices and procedures that properly manage the full data lifecycle needs of an enterprise.

So, to talk of data management, we must start with the beginning, and that is understanding where data comes from.

Understanding Where Data Comes From

As previously mentioned, we generate a lot of data. Pretty much everything you do generates some form of data.

Companies, for example, collect data from internal sources such as transactions, log data, and emails, but also from external sources such as social media, audio sources, as well as photos and videos.

---

[1] https://dama.org/content/body-knowledge

© Stylianos Kampakis 2020
S. Kampakis, *The Decision Maker's Handbook to Data Science*,
https://doi.org/10.1007/978-1-4842-5494-3_2

Other sources of data include

- Published data from credible sources such as government agencies, firms, and industrial associations

- Designed experiments conducted in controlled environments

- Surveys

- Observations

- Automatic, such as user interactions with applications

- Manual entry, which is effectively people inputting information into the system

These sources of data can be further classified by collection methods.

Data Collection Methods

There are two main types of data collection, namely, *observational* and *experimental*.

Observational data collection means that data is gathered *passively* with no attempt to control the variables involved. For example, collecting customer feedback for a book and a retailer analyzing customer behavior are observational collection methods as there is no attempt to control any variable. In fact, the majority of the data collected by companies is done via observational methods.

Experimental data collection involves designing and conducting an experiment where certain variables are controlled while you study other variables. This is most common in academic circles but also in clinical settings.

A perfect example of it is when a pharmaceutical company tests a new drug. They set up experiments where they control certain variables, like the people involved in the study, and they test other variables, such as the effectiveness of the drug and potential side effects.

In business, experimental data gathering is less prevalent. A/B testing is one example of this type of data collection where two variables are tested against each other. For example, a company might test the headline of an article against a different headline to see which is more effective at drawing in traffic.

Data Acquisition Considerations

When it comes to data acquisition, there are certain considerations that you need to take into account. Each one is essential and, as you will see later, they can go so far as to affect your whole business model. These are

1. Appropriateness of the data

2. Nature of the data

3. Time requirement

4. Cost of acquisition

So, let's take a look at each of these and what they mean.

Appropriateness of the Data

If you study machine learning, one of the first sayings you will learn is "garbage in, garbage out." In other words, no matter how amazing your algorithm is, if you don't have the right data, then you aren't going to get the right results.

So, let's say you're a retailer and you want to determine the average transaction size per customer for a particular store, which means you need to analyze the transactions for that store. If you enter the data for the wrong store or input customer feedback instead, then you aren't going to get the results you are looking for. This is a bit of a simplistic and extreme example, but the point is that if you put garbage into your model, the algorithm can't turn it into gold and you will get garbage out the other end.

Now, it might not seem like a big deal, but this happens more often than you realize in the sense that problems are often difficult or impossible to solve because the right data isn't available. This is why it's so essential to keep this point in mind.

Nature of the Data

In some fields, data collection is inherently noisy or biased. For example, consumer surveys and election polling usually deliver biased results. This can be caused by the respondents, but also by the design of the survey, leading to noisy data that's biased from both sides.

Even in more rigid fields, like robotics, there can still be noise when it comes to the measurements made. Of course, this bias and noise can impact the results, which is why it's important to be aware of them. If you don't realize they exist, you won't correct them and the results will be skewed.

Time Requirement

Time requirement refers to how long it takes to collect the data. This is more important in some fields than others, as data collection can be time-consuming. The problem is that in certain situations, it can take so long to collect the data that once you do get it, it might not be useful anymore.

Cost of Acquisition

The cost of acquisition is also an essential component, and it also ties in with the time requirement. The faster you want to get your hands on the data, the more it will cost you. While in some situations the added cost is worth it, in some cases it simply doesn't make sense to pay a lot more to speed up data collection. In some situations, no matter how much you pay, there's no way to make data collection faster.

CASE STUDY: DATA ACQUISITION PROBLEMS

The aforementioned points don't exist in isolation and often interact with each other. The problem is that if you don't take them all into account, you could end up with a business model that doesn't work.

For example, let's say that you've got a brilliant idea for a startup. You want to make money from sports betting by developing a predictive analytics model. While there are companies around already doing this, you want to take it to a whole new level by predicting football outcomes in real time while the game is being played.

So, you want to develop machine learning models to analyze data in real time to determine who will win, how many corners will be won in the second half, and so on. Then, you can bet accordingly and hopefully make a ton of cash.

To achieve this, though, you need data so you check around with vendors to see what pre-defined packages they offer. For each one, you need to consider the aforementioned points.

Thus, you start with *appropriateness of the data*. You have to make sure that the data you are getting from your vendor includes the variables you need to make the appropriate predictions.

The second point is the *nature of the data*. In sports, data mistakes are very common, especially if you want real-time data. The more mistakes you plug into your predictive model, the bigger the chances will be that the forecasted outcomes are wrong.

Third is the factor of *time requirement*. It could take, for example, 10 minutes to get the data after something happened. So, you can't actually get the data in real time or even within a few minutes. Ten minutes might not sound like much, but if you are trying to predict the various outcomes of a game while it is being played, a 10-minute delay could seriously impact the timeliness of your results. After all, getting your prediction after the game is concluded will not help you in any way.

And last but not least is the issue of *cost of acquisition*. To get data faster, you can pay more, but you have to consider whether the additional cost is worth it or not. This is even more important when you consider that the difference in costs can be as much as five times between post-game data and real-time data. Furthermore, real-time data also has an increased chance of introducing errors, which can further magnify the cost of the data by generating skewed results.

When you take the above into account, you have to consider what your next step will be. For example, it might be a better idea to purchase data post game instead of trying to predict outcomes in real time. Not only is this more cost-effective in terms of the actual cost of buying the data, but it also reduces the chances of skewed results. Post-game data is always more accurate because it is double and triple-checked to ensure accuracy.

However, if you do take this route, then you have to consider whether it's still worth getting into this business because there are so many other companies doing the same thing.

And you arrived at this conclusion because you took into account data acquisition considerations, which have shown you that there's a good chance you won't be successful with your idea of real-time sports betting predictions. The advantage is, of course, that you discover these issues before you actually lose money on a model that won't work in the current climate.

Quantitative vs. Qualitative Research

Another important distinction in data acquisition is quantitative and qualitative research. However, these two areas aren't just about how the data is collected but also the type of data and how it is analyzed.

Quantitative Research

Quantitative research is all about numbers. As the name implies, it involves quantifying everything. If you want to know how many people liked a movie, for example, then you turn to quantitative research. If you want to discover how many people buy their groceries online and the frequency with which they do it, then you use quantitative research.

So, quantitative research is used to quantify attitudes, opinions, behaviors, and other defined variables. The advantage is that because of the structured nature of the methods used to collect the data, a larger sample population can be used.

These methods include things like

- Surveys (online, paper, mobile, kiosk surveys)
- Using pre-existing databases (census data)
- Online polls
- Automatic data collection in apps

Quantitative research is great for testing assumptions and getting numbers. However, it's not as good at discovering new insights.

Even when you use open-ended questions in quantitative research, you still can't delve deeper like you can with qualitative research.

Let's say you've designed a new ad and you want to see what people think of it. You might ask an open-ended question like

If you don't like the ad, what is that you don't like specifically?

You'll get lots of different answers from people, but that's pretty much how far you can go. Someone might say they don't like the colors as a simple example. Because of the nature of quantitative research, you have to move on to the next question and can't dig deeper to find out what they dislike about the colors. Sometimes, the insights you can glean from digging really deep can be very enlightening and not always what you thought they would be.

Qualitative Research

Qualitative research doesn't involve numbers or numerical data and is much more free-form. It's all about words and language, but also pictures and photographs. This is the type of research you turn to when you want to understand people's underlying motivations, reasons, and opinions.

Data collection methods tend to be unstructured or semi-structured such as

- Individual interviews or group discussions
- Participation/observation
- Pictures, photographs, videos, and so on

With qualitative research, the sample size tends to be smaller and respondents are selected to fulfill a given quota. Traditionally, it wasn't convenient to go with a larger sample size though things are changing thanks to the Internet.

Before, the most common approach to gathering qualitative data was through interviews and group discussions with a trained interviewer. So, to even be able to come close to the sample size of a quantitative project, you'd need a lot of qualified and experienced personnel, as well as a lot of time, making it unfeasible.

Nowadays, though, there are many other options to gather qualitative data because of the Internet. Thus, this data can be gathered via social media and pretty much through everything your target market posts online, making it easier to use a larger sample size.

DEBUNKING THE MYTH THAT QUANTITATIVE DATA IS BETTER THAN QUALITATIVE DATA

A common assumption is that quantitative research is better than qualitative research because it is quantifiable and can use larger sample sizes. Some also believe this because they think that the structured nature of the collection methods makes the data easier to analyze and less biased.

Nothing could be further from the truth.

First of all, as we've already discussed, sample sizes are no longer as limited as they used to be. Secondly, qualitative data can also be analyzed quite easily nowadays thanks to natural language processing for text data and computer vision for images and video.

For example, you can use deep neural networks to extract emotions from thousands of images and videos or use topic modeling to find the most common process of conversation in a corpus of text (e.g., articles mined from the Web). This process is no longer as subjective as it used to be in the past (when a human had to do all the work), since it is done through machine learning algorithms.

Also, it's very important to note that quantitative data can be still be biased or noisy. Questionnaires with Likert scales or ones that include leading questions, for example, can lead to a glut of quantitative data that is completely biased. Sometimes, respondents to questions deliberately lie. Measurements from devices can be noisy.

Therefore, the lesson is: don't always trust quantitative data, and don't always consider qualitative data to be worthless!

Data Collection Problems

Like with any task, data collection comes with its share of problems that can lead to issues further down the line. While the following list of issues is by no means exhaustive, it does encompass the most common problems that arise in data collection.

First of all, there's the issue of data not being recorded properly. This could be owed to anything from human error when inputting data to noisy sensors, depending on the field. This, of course, causes problems because of the garbage in–garbage out principle.

Another issue is that the data collection isn't supervised by the appropriate principal investigator. When no one who knows the ultimate objective of the project is there to check how and what data is being collected, it can lead to noise and mistakes. If there is monitoring and verification, the potential for mistakes is reduced significantly.

Data isn't always documented, meaning that there are no clear explanations for what each field represents. This makes it difficult for new entrants into the project to understand what's going on and to get up to speed quickly. The delays then slow the whole project down and can even lead to skewed results if the new entrants make the wrong assumptions.

© Stylianos Kampakis 2020
S. Kampakis, *The Decision Maker's Handbook to Data Science*,
https://doi.org/10.1007/978-1-4842-5494-3_3

Then there are the legal requirements involved. In many cases, data isn't held in a way that is compliant with the law, such as GDPR[1] that is huge in the United Kingdom and in Europe. This could potentially cause huge problems later down the line.

Another common problem is when the data is kept only by a specific person rather than being in a centralized system. For example, let's say that the project leader is the only one with the data but his laptop is stolen. The result is all that data is lost, which wouldn't have been an issue if it was stored in a centralized system.

Finally, we get to data being stored in multiple places from different sources, which makes consolidation virtually impossible. This is why an integrated system is so important because it makes consolidation much easier and much cheaper.

Data Collection Examples

Now, let's take a look at some data collection examples. We're going to look at six different areas and how data is collected in each, and we'll also look at the advantages and disadvantages of each approach.

B2C Apps

So, let's start with B2C apps, namely, apps created by businesses for the general consumer. The data collected from these apps is usually from registered users and also from how these users interact with the app.

With data collection via B2C apps, the main advantage is that the data is accurate because no humans are involved, which eliminates the potential for human error.

Of course, there is an exception and that is when users have to register. When users enter their information, there is the potential not just for error but also for users lying. Some users simply don't want to tell you too much about themselves and they have been known to make up things, which can skew your results.

Now, the disadvantage is that the data is only semi-structured, like JSON and XML.[2] The problem is that while the data might follow a pattern, it's not very strictly structured, like having it in a table format. This creates an issue for data scientists because it's not always easy to figure out how to manipulate the data or even what variables should be collected.

---

[1] https://eugdpr.org/
[2] The W3 schools site has a great tutorial on JSON www.w3schools.com/js/js_json_syntax.asp and XML www.w3schools.com/xml/

For example, a data scientist would have to figure out if every button press and swipe is relevant and should be collected. This is important because the more data you collect, the longer it takes to wrangle it into something that can be analyzed and converted into valuable insights. It can take days or even weeks to get the data under control and translate it into something useful.

Sales

In sales, data is often collected using a CRM, that is, a customer relationship management tool. While some companies might have a different type of tool—or they at least call it something different—in essence, it's still the same thing. In other words, it's a system where customer data is recorded.

There's also company data obtained from various databases or other sources, such as UK Companies House or databases like Orbis.[3]

The advantage to sales data is that it is generally easy to interpret. You have things like number of units sold, number of calls, average transaction value, and so on. It's pretty straightforward.

However, the problem is that it isn't always clear how the data can be used to provide answers to questions. For example, in your CRM you might have notes on each client, and it might not always be clear how to turn those notes into something useful. It's not impossible by any stretch of the imagination, but it isn't always as straightforward as it might seem.

Retail

Then we move on to retail, where you have sales data—essentially data collected from transactions—as well as customer data.

The main advantage is, like with the aforementioned sales data, the data is pretty straightforward and easy to interpret. You have number of units sold, transaction values, and so on.

However, there are legal considerations to take into account with retail data. For a retailer, the more data they have, the better. But it can be difficult to identify it down to the individual. And this can be, in large part, due to legal frameworks that limit the way retailers can use the data they collect.

---

[3] www.bvdinfo.com/en-us/our-products/company-information/international-products/orbis

Finance

In finance, there's stock market data. This is probably the best area for data collection because the process is automated, the data is clean, and it comes fast. It's one of the few areas where it's virtually impossible to find problems with data collection.

Sports

Then there's data collected in sports, which can be player data that is obtained using an external provider, such as OPTA[4] or STATS.[5]

While sports data is structured and relatively easy to interpret, the main problem is that it is entered manually. In other words, people sit and watch the game and then record the data. This, of course, can result in mistakes and often does as humans are prone to error.

Social Media

Last but not least, we have social media.[6] As you can imagine, this is data sourced from platforms like Facebook, Twitter, Instagram, Pinterest, and so on.

The advantage to social media is that data collection is automated. All you have to do, basically, is plug in an API and off you go. However, the problem arises when you try to make sense of all the data you collected. It's completely unstructured and really, really noisy and messy.

If you've ever worked with any form of social media data, like Facebook posts, tweets, and even blog comments, you've likely seen a wide range of issues. You have grammar issues, different languages, special characters and emoticons, and so on. As you can imagine, this can make it very difficult to wrangle the data into something useful and easily analyzed. It definitely takes a lot of time to do so.

Data Management Practices

Now, let's take a closer look at data management. There are both good and bad practices in data collection and storage, and we'll be looking at some of these.

---

[4] www.optasports.com/
[5] www.stats.com/
[6] Read more about social media analytics here: http://thedatascientist.com/social-media-analytics-buzz/

However, it's important to note that we aren't going to be looking at good and bad practices from a technical standpoint as this isn't a book about data engineering. We'll be analyzing these practices from a business viewpoint, and once we're done, we'll also take a look at some examples of bad practices and good practices.

Good Practices for Data Collection and Organization

Good practices in data collection and organization mainly revolve around having an objective before you start, being aware of how data collection affects the rest of your business, and creating a standard. Let's examine each one of these elements.

Establish a Goal First

When it comes to data collection and organization from a business perspective, it is essential to first establish at least one goal or have one question to answer. Otherwise, you will end up collecting data that you might not need or the wrong data.

So, before you do anything, you need to establish an objective. Before you ask, "let's collect some data because that's what everyone seems to be doing" is not an objective. On the other hand, "why do 35% of our customers not go through with the purchases in their cart?" is a good objective/question to start with.

Awareness of How Data Collection Affects the Rest of Your Business

It's also essential to be aware of how data collection interacts with other parts of the business, such as the user interface. We'll be talking more about this later.

Establish a Data Standard

You also want to use a data standard and document everything that is happening. By creating a data standard everyone can easily understand, new project entrants can quickly get up to speed, saving you significant amounts of time and resources in the long run.

Bad Practices for Data Collection and Organization

Bad practices are pretty much the exact opposite of the good practices I described earlier. Let's take a closer look at each.

No Clear Objective

Having no clear objective from the start means you just collect data for the sake of it with no clear direction and take the "we'll store the data now and think about it later" approach. This is pretty much a waste of time and resources.

Ignoring the Connection Between Data Collection and the Rest of Your Business

The next issue is not taking into account the connection between data and business. You might think that the data won't affect the business at all, but this is completely erroneous. For example, if you want to build a better recommender system and to collect data to this end, you might have to completely rebuild the UI. Nothing in your business exists in a vacuum, including and especially data collection.

No Documentation or Data Standard

Finally, the most common bad practice is having no documentation or data standard in place, which makes it difficult for anyone new to the project to understand what's going on.

Examples of Bad Practices or What You Shouldn't Do!

In this first example, we'll take a look at how lacking a clear objective can derail your efforts and cost you money.

Lack of Clear Objective

Let's take a B2C app that helps users find restaurants in their country or city as an example. You run and manage this app and you decide to collect data from users. You have no idea what you're going to use the data for, but you figure that it will be useful at some point in the future.

You decide to register the information in a NoSQL database in JSON format. You collect their name and age, like {"name": John, "age": 23}. However, you decide to collect the data generated by user interactions in the same format, such as { {"21-10-2017":{"opened_the_app":"yes", "paid_through_the_app":"yes"}}}.

Thus, you now have their name and age, and you also have their app interactions, like when they open the app, scroll through it, make a selection, and so on.

One day, a year after you launched the B2C app, you decide you should find a data scientist to help you with the data. So, you hand over the data you've collected to the data scientist and ask them to extract general insights.

You might be wondering what's wrong with this approach—it doesn't seem to be problematic, does it? Well, there are actually quite a few issues.

This is one of the most frustrating situations to be in when someone is a data scientist. Why is that? Well, first off, you haven't optimized the data to achieve any particular goal, which means that data scientist will waste a lot of time trying to figure out what they can do with the data and how they can do it.

The data wrangling will also take a really, really long time. And this will be time that costs you money, whether you are outsourcing to an independent data scientists or have hired a full-time person for the position.

Finally, you opted to use NoSQL as the format for data storage, and this might not necessarily be the best option depending on what you want to do with the data. So, this will make it even harder for the data scientist to wrangle the data and it will certainly take a lot longer.

But I Have an Objective!

Okay, so let's take the same B2C app from the previous example and assume you have an objective. That should make it easier to wrangle all that data into something useful, right? Well, not really.

Let's say that your goal is to build a recommender system so you hand the data over to the data scientist who should be pleased that you have a specific goal you want to achieve. What's wrong now?

While it's certainly a little better than the previous example because you do have an objective you want to achieve, issues will still arise. First of all, the data might not be in a format that is optimized to develop a recommender system. This means that the data scientist will have to try to wrangle the data into submission to get it in a format that is usable.

Furthermore, if you collected data willy-nilly without knowing the objective in advance, you might not even have the right variables available. You have age, name, and interactions, but what about gender? Does gender even matter in recommending a restaurant? Of course it does.

Here's a really extreme example. Say, you have a number of "restaurants" registered that provide adult entertainment for women, a.k.a. male strippers. . How happy would you be if you were searching for a family-friendly restaurant and saw a strip club "restaurant" in the list of recommendations? Not exactly overly pleased, right?

And yes, we know that a lot of other variables would be important in the previous example, but it just goes to further prove the point that you need to collect a lot more data to create an effective recommender system.

Another variable you'd definitely need is location. What's the point of recommending a restaurant in New York to someone who lives in Austin? A recommendation like that isn't of much use, and it will also alienate your users because they'll quickly realize your recommender system is a complete waste of time.

Yes, a number of these variables can be inferred, but it is very time-consuming and prone to errors. So, it would have been a much better option to collect this information from the start.

The conclusion is that a little advanced planning can save you lots of pain in the long run. Remember, as with everything, failing to plan is planning to fail when it comes to data collection.

CASE STUDY: BAD PRACTICES IN DATA MANAGEMENT

Some of the things discussed in this chapter might seem too obvious. However, the kind of problem you can encounter in your daily professional life might surprise. Here's an example from personal experience—just in case you think stuff like this doesn't happen in the real world.

A few years ago, a client wanted me to study the difference between male and female behavior on their platform. Do males take more actions X than females? Or do females consume more than males on the platform?

This is a relatively straightforward problem which can be answered through simple statistical means such as hypothesis tests. However, things were not so simple as it would have been expected.

Gender wasn't a variable that was collected and stored in the database. So, the objective of the study was to do understand the differences between genders, but this variable was non-existent!

The solution that was devised was to infer gender from the names of the users. However, in a multicultural place such as London (where this company was based), this is much harder than it sounds. There are so many nationalities living in London, which made it impossible to cover all the names by using a simple name registry.

Unfortunately, even after using name registries from multiple countries (ranging from the United Kingdom, to the United States, to Nigeria, to India, and more) we still had to throw 30% of the data. The reason was that some names could not be found in the registries and that some names are used by both males and females.

Therefore, what should have been a straightforward piece of research suddenly turned into something very complicated. And the root of all these causes was the lack of planning and the lack of a data strategy. *Failing to plan is planning to fail.*

So, whatever you do, do *not* be like this client. Plan your needs in advance so you don't waste your time and resources or end up with unusable data and skewed results.

Examples of Good Practice or What You Should Do!

This is the perfect time to discuss what you should do. Coming back to our restaurant B2C app, it would have made life easier for everyone if the data strategy had been established from the beginning.

So, whatever app or business you run, you need to think in advance of what you might need. With the restaurant B2C app, it would have been clear from the start that a recommender system would be necessary at some point. And the same holds true for most of these apps.

If you have no idea what you might need in the future, you can just hire a data scientist to consult. It might cost you a few hours of consultation, but it's going to save you a significant amount of pain down the line. It'll also definitely be cheaper because setting everything up right from the start will always be cheaper than fixing the problems in the future.

Another issue is with the technology being used. Why choose NoSQL if for certain things SQL is much easier to deploy and use, like for user registration data? It's like using a grenade to kill a fly instead of a fly swatter or fly spray. It might get the job done but it's definitely overkill and will create a much bigger mess.

Finally, you need to be aware of all the variables you might need in the future. Storage is cheap and it's better to have more information than you need rather than omitting data and not recording variables that might be critical to future projects.

Obviously, you can't think of everything in advance, but it's still a good exercise to try to do as much as possible.

Don't Buy the Hype!

This is a very important point. In computer science, new technology is being developed constantly, and something new and shiny pops up every few years. For example, for big data, we saw the development of NoSQL databases, while machine learning gave us deep learning.

While these technologies are extremely powerful, they've been created to deal with specific issues. The problem is that there are companies out there that want you to buy their brand new shiny products and services and will say just about anything to convince you that you need them, even when you don't need tech that advanced.

This is quite a common issue. There are so many companies out there using NoSQL solutions, for example, when they don't really need it. Of course, this doesn't mean you should never use these technologies. It just means that you should think about it twice and make sure you really do need them.

For example, NoSQL generally assumes that you want to store big amounts of data and it's difficult to express the structure in a table format. So, if you don't have ridiculous amounts of data, it might not make sense to go with NoSQL. Also, choosing the right NoSQL database is not an easy task. Just by searching the Internet,[7] you can find more than ten different options, with more coming. There are some cases where using a NoSQL solution makes perfect sense (e.g., graph data or when you are storing posts from social media), but don't just do it without thinking. In most cases you are just fine with some simple implementation of SQL, such as MySQL.[8]

Deep learning is another example. It's great technology but should only be applied in certain situations, like for images, audio, and video. Otherwise, it requires a lot of computational resources, which means it can get expensive and time-consuming to develop a deep neural network.

In other words, don't use technology just because it's new and shiny and a slick sales rep has spent an inordinate amount time convincing you that you absolutely need it. Instead, take a step back and take an objective look at if you really need that type of tech.

If you don't, it's better to go with a trusted solution that has been available for a longer period of time because it's been thoroughly tested and proven to work effectively. Plus, it will also be easier to hire people to work with said technology because there are more people who have expertise working with it.

---

[7] https://en.wikipedia.org/wiki/NoSQL#Types_and_examples_of_NoSQL_databases
[8] www.mysql.com/

Setting Goals in Advance

Failing to plan is planning to fail. This is one of my favorite phrases when it comes to data strategy.

If I haven't yet made my point, I'll say it again: it's essential to think about what you might want to do with the data in the future.

It's definitely not easy to think in advance of all the types of data you might need to solve a problem, and it can take time. However, it more than pays off to sit down and think about it.

For example, you might need 20 variables to create an outstanding recommender system for your app. If you didn't take time to think about it, you might have 3 of these variables. Conversely, if you did sit down and plan in advance, you might have 15 variables. It's not the full 20 because no one can predict everything in advance, but it's still much better than only 3.

If you really have no clue or don't have the time to engage in this exercise, then hire a data scientist to consult before you start collecting data. Taking this approach is also essential because data collection and data science can interact with other parts of your business and business model. It's not going to take more than a couple of days of the data scientist's time (which means you won't have to spend much money), and it is well worth it.

Examples of the Impact of Data Science and Collection on Your Business

In this section, we're going to look at some examples of how data collection and data science can affect your user interface and can even lead to completely different business models.

Data Science and Dating

So, let's say you have a B2C app in the dating space and you need to create a recommender system[9] for your app.

Consider Tinder, a popular dating app, and OkCupid, which is a popular dating web site. It's not important if you aren't familiar with these apps, but they are excellent examples.

From a data science or machine learning perspective, the whole problem with dating is ensuring the system makes the right recommendations.

---

[9] More information about recommender systems: http://thedatascientist.com/right-way-recommender-system-startup/

Tinder, for example, takes a pretty simplistic approach. You can swipe left or right on someone's profile, which tells the system whether you like them or not. However, you don't have a lot of data. You can see their picture, first name, and age, and maybe a few other photos, but that's about it.

OkCupid, on the other hand, has taken a much more in-depth approach. They've created a questionnaire that has a lot of questions. And when I say a lot, I really mean a lot of questions. This questionnaire really digs into your preferences, opinions, and personality. They then use the answers from the questionnaire to match people with each other.

If you're a data scientist and you want to create a recommender system for matchmaking, you're probably going to do something similar to what OkCupid has done.

However, you have to consider the fact that OkCupid is much more difficult to use. The site asks you over a hundred questions, and it's going to take you a while to get through them. So, in terms of data science, OkCupid is a data scientist's dream. The data that OkCupid possesses is so detailed that they have even published research on human sexuality and romantic preferences.[10] However, this does not necessarily make for the smoothest user experience.

Conversely, Tinder might never have a great recommender system because they don't collect enough data to develop one, but they do provide a great user experience. That's where their focus has been and it can be seen quite easily because it takes very little time for users to screen a large number of profiles.

One is not necessarily better than the other. They're just different. But what is important is that these two systems clearly show how data science and data collection affect the user interface, the user experience, and even the whole business model.

And it should be noted that both OkCupid and Tinder are wildly successful. So, it's not a matter of which one is right; it's simply a matter of you taking the time to think about how this works in your own business.

Data Science and Entertainment

Another example is of B2C apps in the entertainment space, more specifically concert finding apps. Recommender systems make it easy to demonstrate how all the different elements interact, which is why we are focusing on them.

---

[10] https://theblog.okcupid.com/a-digital-decade-sex-c95e6fb6296b

With a concert finder, you can scroll through the list of local shows. You tap on a specific show; you have a number of options. You can state that you're attending, you can say maybe you're attending, you can get tickets, or you can share on Facebook or Twitter.

From a user perspective, this might not seem like important information. But for a data scientist, it's a wealth of data—it's like hitting the jackpot.

Every action demonstrates a different kind of involvement. Someone who clicks "I'm attending" is definitely more interested and involved than someone who only says "Maybe." And then there is the combination of actions.

Someone might state they are attending and then also share the concert on Facebook, which clearly shows that they really, really love that music and/or performer.

So, the important part is that all this data is stored in a database and can be analyzed to garner important and valuable insights. A data scientist will be able to see all these actions and combination of actions and draw conclusions that will benefit how the recommender system works.

For example, if someone states they will be attending a Beyoncé concert, gets tickets, and shares on both Facebook and Twitter, then they're going to want to know about every possible Beyoncé concert—in some cases even if it's not a local concert.

However, someone else who simply says they might attend and eventually does get tickets but doesn't share might not be as interested. While you will definitely include local Beyoncé concerts in their recommended list, it might not make sense to tell them about concerts outside their area because they aren't such an avid fan.

So, this information could help a data scientist understand how important the different actions are, for example. This can, of course, be used to later improve the recommendations the system provides.

It can also be used to understand how to convert those who said "Maybe" into ticket buyers. A company could send them offers or rewards or use different methods to change their mind.

This shows how certain aspects of a UI might mean very little to the end user but can be a treasure chest of vital information for a data scientist and can have a significant impact on the success of your business.

How to Keep Data Tidy

Dirty data refers to data that can suffer from all sorts of problems, including, but not limited to, things such as erroneous or conflicting entries, missing values, and outdated data. Tidy data is the opposite, data that is in a nice format, with no inconsistencies or other issues.

Dirty data can cause all sorts of problems. First, it makes consolidation of different data sources difficult or sometimes outright impossible. Second, many of the data points might not be usable. This can reduce the effective size of your data. You might be holding 5GB of data, but only 1 gigabyte might be usable.

There is another tricky problem that can show up with dirty data. If you create models based on dirty data, then the model will produce results that might not correspond to reality, but you might not be aware that this is happening until it is too late.

For example, you might have deployed a machine learning system in the real world (e.g., a recommender system, or a price optimization tool, or a demand forecasting tool), and it might take you some time before you realize the results do not correspond to reality. However, in the meantime, you have been recommending products (or adjusting prices, or something else) in the wrong way!

© Stylianos Kampakis 2020
S. Kampakis, *The Decision Maker's Handbook to Data Science,*
https://doi.org/10.1007/978-1-4842-5494-3_4

This is one of the trickiest parts behind data science models. A wrong model will not fail to compile (like in software development). It will work fine until you realize that the results do not make much sense. This is why planning in advance is so crucial (as we stressed out in the previous section), and keeping your data tidy is part of this plan which you must follow.

In terms of causes, most problems with dirty data are usually caused by

1. Different departments handling the datasets

2. Lack of documentation

3. Lack of communication between different parts of the company

The easiest way to understand dirty data and its potential impact on your organization is to look at a real-world example. This actually did happen like all the examples woven throughout the book.

Let's say you are dealing with data collection and data management in a football club (or soccer club if you're from North America). So, within a football club, you have a medical department, football training, weight training, and the coach.

However, there is no centralized system for collecting or storing the data. All data is collected manually and stored in Excel files, which is very problematic. It's very difficult to format, use, and consolidate the data.

Different people have different files and different datasets. Essentially, the data is all over the place and everyone has bits of it.

There's also a lack of supervision, which means that there is no certainty the variables recorded are accurate.

Then you have a limited communication between departments. For example, a player might come back to train after suffering an injury, but might not let the medical department he's back. So, the medical department believes he's still injured and has him recorded as such in their data.

The departments also have different goals. The medical department wants to protect the player to ensure he is fully healed and won't end up even more hurt if he comes back too soon. Conversely, the coach just wants the player back on the field. So, the medical department could have an incentive to make the injury seem more serious than it actually is, while the coach wants to minimize the importance of the injury.

Lastly, there's no established procedures for data standardization and data entry. This means there's a certain level of freedom in terms of how the data is being recorded by each department. It can even occur within the same department. Two different doctors, for example, might refer to the same medical issue in two different ways. For a doctor, it might be easy to understand that they're talking about the same thing, but for someone who isn't a doctor, it can be a problem. So, a data scientist trying to analyze this data could end up confused.

There's also lack of transparency. In our case, the medical department had no idea what was happening in the player's schedule. It's not just about the player who didn't inform the medical department he's recovered from his injury.

For example, a player could be eating the wrong food, and no one is recording it. Likewise, a player could have been training longer hours without informing them.

Of course, without this vital information, it makes it harder for the medical department to resolve issues and diagnose problems. So, our player with the poor diet could end up having issues with performance, but the medical department can't advise him on a better diet or at least the effects of his current diet on him before it becomes a performance issue because they don't know.

Thus, they ended up with data that was "dirty," which resulted in lots of back-and-forth. A lot of time was spent in meetings between the different departments, trying to sort all the data out before anyone could analyze it properly.

In the meantime, a lot of time was wasted as it took months to clean up and consolidate the data, but worse, a large number of erroneous conclusions were drawn.

Finally, the learning curve was long. First, the mess meant that the overall time of the project to make sense of the data was long. Second, any new people who joined the project took quite a while to get up to speed and to reach a point where they understood what was happening.

Solutions

So, the first solution is to implement a centralized system to store data, but also a single system to collect data. In other words, Excel shouldn't even be a blip on anyone's radar. Furthermore, no more separate files, with each person having their own file on their own computer in their own format.

Now, there's a centralized data storage system, and all data is collected through a single app that everyone uses, with no one recording data manually anymore. This approach has made it easier to enforce a data standard.

Thus, people no longer had that much freedom in how they recorded data. For example, doctors couldn't describe the same medical issue in different ways because they had a limited number of options available in a drop-down list. This minimized the possibility of misunderstandings taking place.

It was also essential to improve communication between the departments. Thus, quick weekly meetings were instituting, essentially "forcing" departments to talk to one another. This helped eliminate issues where a player was actually playing while the medical department still had him recorded as being on leave due to his injury.

So, the solution involved creating a data standard, forcing people to follow the standard, centralize everything, and essentially eliminate silos. When you have silos, it's very easy for misunderstandings to occur, so when you eliminate them, you will find that efficiency increases significantly.

CASE STUDY: MODELS ON DIRTY DATA

This true story can help you understand how dirty data can mess things up. While doing research with a football club, I thought I had discovered a very interesting pattern. Players that were training during the pre-season period seemed to get injured on Mondays after a couple of weeks of training.

I create a statistical model which seemed to validate this conclusion. The working theory was that there must be some kind of shock effect into the body coming back from the off-season and straight into intensive training during the pre-season. Maybe starting training and then being off training for the weekend and coming back to it on Monday increased the likelihood of injury. If that was the case, then the club would have to adjust its training methods or resting protocols for the players.

However, my theories were soon confronted with reality. The explanation was much simpler. There was nothing happening with overtraining or other complicated patterns of the athletes being active and inactive which made them more prone to injury.

In many cases injuries would not be detected by the coach or the training staff. Rather, the players had to report the injuries. The team's protocol dictated that if a player reported being injured toward the end of the week, they had to stay in the training facilities for rehabilitation for the whole weekend. However, most players didn't like that. They wanted to just go to a pub or stay at home or whatever it is that they preferred to do with their free time on a weekend.

This means that many players that were injured late on Thursday or on Friday would not report the injury on those dates. Rather, they would wait for Monday and they would report they are injured. Hence, the pattern I thought I had discovered had just been caused by the peculiarities of the domain and data collection. Injuries were not objectively measured (rather self-reported). Injuries had been recorded as taking place as soon as they were being reported. For example, if someone was injured on Friday, but reported it on Monday, then the injury was recorded as having taking place on Monday.

A couple of months of work down the drain because of wrong data collection practices and gaps in domain knowledge caused by miscommunication. This is a case where dirty data really messed up the whole plan. Lessons learned? First, when entering a new domain, ask as *many questions as possible*, as the domain experts will not always remember every relevant detail and explain it to you. Second, make sure you fully understand the data collection and data entry practices. Again, these problems would have been avoided if a proper protocol had been designed from the beginning.

Thinking like a Data Scientist (Without Being One)

In this chapter, we're going to talk about how to think like a data scientist without being one. It's probably the most exciting and interesting chapter of the book, but also the most important because it translates very technical terms and ideas into English. It offers a great view into the technical world of data science but without all the jargon, so you can understand what's happening.

© Stylianos Kampakis 2020
S. Kampakis, *The Decision Maker's Handbook to Data Science*,
https://doi.org/10.1007/978-1-4842-5494-3_5

The Data Science Process

So, if I ask the question

What is data science about?

I'll receive different answers because, no surprise, different people have different opinions. For example, the boss might think that data science is actually a magical money machine that can print money whenever he or she wants it to.

To customers, data science might make you seem like a wizard or a telepath or psychic because you can tell so much about them and solve so many of their problems.

Software engineers think that data science is just importing some stuff from libraries and running a few functions, because of course no one is quite as awesome as they are.

In reality, though, data science is like an orchestra made up of different elements, including infrastructure, software, data sources, and statistics.

From a pragmatic point of view, there are five steps of a data science life cycle:

1. Collecting data
2. Organizing data
3. Analyzing data
4. Interpreting data
5. Communicating findings from data

We already covered collecting and organization data in the previous chapters; therefore, we're going to look at steps 3–5 in this chapter and we'll cover the topics in enough depth so that you understand how it all works.

Keep in mind, though, that data scientists are often called in from step 3 to step 5, because the collection and organization has already been done. Now, that doesn't mean it's always been done properly, but it has been done.

Defining the Data Science Process

The data science process has four steps, namely:

Step 1—Defining the problem

Step 2—Choosing the right data

Step 3—Solving the problem

Step 4—Creating value through actionable insights

The process involves two main actors, namely, the **domain expert** and the **data scientist**.

The *domain expert* is the person who owns the problem. It could be the business owner who hired the data scientist, but it could also be the head of a department that has to work with the data scientist. In an academic setting, it could be a researcher.

Essentially, the domain expert is the person who is responsible for solving the problem and who has to report either to himself (as a business owner or entrepreneur) or to upper management. It's also the person who knows all the specifics of the domain.

Thus, the first step, namely, defining the problem, lies with the domain expert. Subsequently, choosing the right data is the responsibility of both actors, while the third step falls to the data scientist. Finally, creating value involves both actors working together.

So, let's take a closer look at all of these.

Step 1: Defining the Problem

The domain expert is in charge of defining the problem because he or she is the one who understands the domain better than anyone else. The issue is that domain experts often don't know data science, which can create problems.

And that's exactly what this section is about. It's about understanding why you need data science, how you can use it, and under what circumstances it becomes relevant.

If you have a data scientist in your company who has been working with you for a long time, they likely have a good knowledge of the domain. This means they can define their own problems.

However, in most cases, it's the domain expert who lays out the problem. They might want to forecast sales for the month, or they might say that the recommender system needs to be improved, for example.

Once the problem has been defined, it's time to move on to the next step.

Step 2: Choosing the Right Data

The domain expert needs to understand *data collection* and *management*, which we discussed in the previous section. He or she is responsible for making sure the principles on data collection and management we covered previously are being followed so that the data is as clean as possible, the standards are documented, and the data is easy to use.

The data scientist, on the other hand, needs to understand the domain and any peculiarities it might have. We touched on this in the data management section too.

So, each domain has its own set of pros and cons. Thus, in some cases, you can have very noisy data, like if you work with sensor data. In other cases, data might be inaccurate, such as the data collected through polling and surveys as people might not be completely honest in their answers. And the data scientist needs to know these things.

Step 3: Solving the Problem

The domain expert's main contribution to this step is ensuring they hire the right person, which we will discuss in the next section of the book. They're also responsible for ensuring that the company has the right culture, which we will cover in the last chapter of the book.

The actual solving of the problem is the data scientist's job, and there is little that the domain expert can do other than provide support.

Step 4: Creating Value Through Actionable Insights

This step is the responsibility of both the data scientist and the domain expert. It requires a good collaboration between the two parties because the domain expert needs to know how to work with the data scientist and also to understand what the result of the work entails.

At the same time, the data scientist also needs to understand by now some of the issues of the business and of the particular field.

At this point, it really pays off if the person who is the domain expert, such as the business owner, the manager, or the decision maker, has a good understanding of data science.

It also helps build a *data science culture*, but we'll cover this more in depth in the final section. The right culture can make it easier for the data scientist to learn the peculiarities of the field and the issues of the business. This leads to a data scientist who can take a more creative approach to developing solutions to help the business.

Solving a Problem Using the Data Science Process

We're going to look at an example. It's a very simple issue, but it is the best way to show how a problem can be solved using the data science process.

Step 1: Defining the Problem

So, the first step is to define the problem. Take something you find interesting at work. You could even decide to focus on something that bothers you.

I'm going to choose a trivial example, but one that most people can relate to, namely, meetings that consistently start late.

The first thing to do is to frame the problem. A great trick is to turn the statement of the problem into a question and write it down. So, in our case, it would be

Meetings always seem to start late. Is that really true?

You want to take this approach because while most data scientists like concrete terms, most non-data scientists tend to think in terms that are vaguer. However, models, be they machine learning, statistical, or any other type, have a narrow focus and can answer one, two, or three specific questions. So, when you write down the problem as a question, it helps the data scientist develop a better model to reach the solution.

Step 2: Choosing the Right Data

This step will have two parts, namely, Part A where you think about the data and Part B where you collect the data.

Part A: Think About the Data

At this point, you need to consider what data you will require to answer the question you framed in Step 1. And then you have to decide how you will collect the relevant data.

You start by writing down all the relevant definitions and thinking about the protocol you will use to collect the data. This is when the problems start, and you realize that even for simple problems there might be a wide range of different issues.

So, in our example, we want to study meetings that start late. To do this, though, we need to define when the meeting actually starts. It might seem simple at first, but if you really think about it, the definition isn't all that clear.

Thus, does the meeting start when someone says it's time to get started? Or is it the time the meeting was scheduled for in the calendar? Or is it when the actual work of the meeting starts—you know, after all the chitchat and enquiring about the boss' kids, dog, cat, hamster, and crocodile? Or is all that small talk that takes place in the first few minutes part of the meeting too? Some might think that bit is a waste of time, while others might find it important for relationship building.

It might seem pretty trivial, but all these things are important because things can go wrong. The main risk is that you use the wrong definitions and they prevent you from solving the problem.

For example, meetings might consistently start late because of small talk. However, when you wrote down the definition, you didn't factor in this small talk and, therefore, recorded a different time for the start of the meeting. So, you can't solve the problem because the cause is completely missing from the data. As a result, you have to go back and collect data again, thereby wasting time, energy, and resources.

Part B: Collect the Data

Now that you've defined your data, it's time to move on to the collection part of the project. The first thing to consider is that you need data you can trust, which can be more difficult than it sounds.

Data can suffer from a wide range of issues such as missing values, erroneous entries, definition issues, and much more. And these are pretty frequent problems. We covered this in the previous section as well as how to prevent or fix these problems.

Due to these problems, it's a good idea to be flexible and adaptable. For example, it could be a good idea to modify your definition and the data collection protocol as you go along if things aren't progressing as they should.

So, come back to our meeting example. Let's say your initial definition doesn't include the "small talk" portion of the meeting. However, you realize it's extremely relevant since it delays the start of the actual meeting by approximately 30 minutes. Well, you can't solve the problem if the data doesn't show it exists, so you go back and change the definition but also how you collect the data.

What is vital, though, is that there is complete transparency and everything is documented regarding what you changed, how you changed, when you changed it, and so on. This is important because you could end up with situations like variables meaning completely different things but using the same terms to define them.

For example, let's say you don't document your decision to include small talk as part of the meeting. The time you record as the start time changes, but no one knows about it because you didn't document the change.

Then, you comb through the data with the data scientist, who can't figure out why nothing works properly.

The reason is, of course, that the variable referring to the start time has two different definitions depending on when it was recorded during the project, namely, before or after you made the change.

Effectively, you're dealing with two variables and not one, which can seriously mess up any statistical model.

This is just one example of how things can go wrong when you don't document everything you are doing. Many more things can occur, leading to disastrous results.

As you aren't a data science expert, there's no way for you to foresee what problems might arise with the data, which is why you want to be fully transparent regarding what's going on and to document every tiny little thing that you do.

Remember, you always want to make the data scientist's life *as easy as possible* because it will take them less time to get up to speed and it will enable him or her to develop a more effective solution, helping you save time and money while greatly improving efficiency and productivity.

Step 3: Solving the Problem

In this phase, there's not a lot you can do as the domain expert since it's mainly the data scientist's job. However, let's take a quick look at what this step usually entails anyway just so you can get an idea of what's going on.

Usually, most data scientists like to start by performing exploratory data analysis. They use summary metrics, graphs, and so on to gain a better understanding of the problem. In some cases, they might even present some early findings to the stakeholders.

So, in our example, a data scientist might say something like:

"Over a two-week period, 10% of the meetings started on time. On average, meetings started 12 minutes late."

Or, he or she might discover that meetings usually started late when they covered topics X and Y and when certain people attended, like the manager who loves to tell everyone what he did over the weekend in minute detail and no one can seem to get him to stop talking.

The data scientist will then build a model to find a solution, but models are mainly used to answer specific questions. You can have a primary question, which in our example could be

Are people usually late for meetings?

But you can also have a number of secondary ones. In our example, they could be along the lines of

- Does the subject being covered affect how long a meeting runs?

- Do the people in attendance have an effect on how late the meeting is?

- Are late meetings counterbalanced by meetings that finish early?

- What is the overall meeting time? Can you describe this in terms of a distribution?

Of course, all these questions are related and that means more than one can be answered. Keep in mind, though, that when you frame your problem as a question, you are helping the data scientist. He or she will be better able to understand how to convert vague data into a model that can be used to solve the problem.

Step 4: Creating Value Through Actionable Insights

By this point, the data scientist has built one or more models to answer the aforementioned questions, so we can move on to the final step, namely, to extract actionable insights from the models.

Essentially, the models are useless if you don't act on them. The first step to taking action is to understand the impact of the results. A good way to do that is to translate the findings into monetary terms.

In other words, if the time period studied for late meetings is typical, then each employee loses an hour per day, which costs the company $X per year.

Another approach to understanding the impact of the results is by identifying specific individuals or culprits behind the problem.

So, in our example, we could say that person X is responsible for late meetings, so what can we do to solve the issue? Could we help them manage their time better? Would that help fix the problem?

Based on this understanding you gain from the results, you can then take action.

It's also a good idea to go beyond the results, though. This cycle of modeling, translating the results, and then taking action can also be a very good learning experience for you. When you go through this process, other questions pop up. This is the point when some of those secondary questions mentioned in step 3 arise.

It also makes sense to build models for these secondary questions too since the data is already there and the data scientist has gained experience with the data. It pays to make the most out of the data.

A Short Introduction to Statistics

Statistics has two branches, namely, descriptive statistics and inferential statistics (Figure 6-1).

© Stylianos Kampakis 2020
S. Kampakis, *The Decision Maker's Handbook to Data Science*,
https://doi.org/10.1007/978-1-4842-5494-3_6

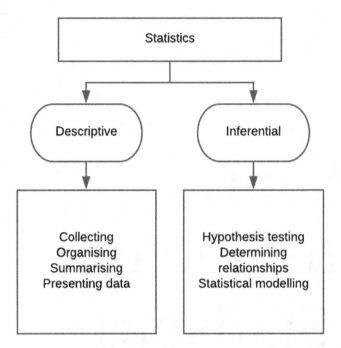

Figure 6-1. The two branches of the science of statistics

Descriptive statistics is what most people associate with the word **statistics**. These are things that you've probably done in high school. They include things like collecting data, using summary metrics such as the mean, and visualizing data.

Inferential statistics is what most statisticians think of when they think of the word **statistics**. It involves more advanced concepts such as sampling and inferring the values of the parameters of the population.

Descriptive Statistics

Descriptive statistics deals with organizing, summarizing, and presenting data in a convenient and informative way. Two sets of methods are used in this type of statistics.

First, there are numerical methods for describing the data. These include measures for the location of the data, like mean, median, and mode. They also include measures of the variability of the data, such as the range, variance, and standard deviation.

The second set of methods cover graphical means of presenting data, which you probably know quite well, such as bar charts, pie charts, histograms, and so on.

I like to refer to descriptive statistics as "high school statistics." The reason is that most people are taught descriptive statistics in high school and they think that descriptive statistics are representative of the overall field of statistical science. However, the truth is much different.

Inferential Statistics

When a statistician is talking about statistics, this is usually what they refer to. Inferential statistics is much more interesting and useful than descriptive statistics. It basically represents 90% of statistics, with descriptive statistics representing merely the tip of the iceberg.

Thus, inferential statistics is a body of methods used to draw conclusions or inferences about the characteristics of a population based on sample data.

Some of the fundamental terminology used in inferential statistics includes

- Data, elements, variables, and observations
- Population—The group of all the items a statistician is interested in
- Sample—A set of data drawn from the population
- Parameter—A descriptive measure of a population
- Statistic—A descriptive measure of a sample
- Inference—A set of methods, such as estimation, testing, and prediction, used to draw conclusions or inferences about the traits of a population based on data from a sample

So, what does this all mean? What does inferential statistics actually do?

Inferential statistics allows us to quantify uncertainty. In other words, let's say you want to find out something about a population. Let's go with something simple, like the average height of the people in a country.

Now, to determine this figure, the first thought would be that you have to take every person in that country to determine their height. Or, measure every person. When a country has a population of 50 million, that isn't going to be an easy undertaking. Even for something so simple, you are going to be spending a lot of time and money trying to talk to every person.

Studying every individual is, at the very least, impractical, but in most cases impossible.

With inferential statistics, you can take a sample of the population and express uncertainty through a probability distribution. It helps us turn the uncertainty into a quantifiable metric, which then allows us to run calculations. This is what allows us to do statistical modeling and hypothesis testing.

So, essentially, what happens is that you can take a sample of 1000 people, for example, and extrapolate their average height to the entire population of 50 million. And working with a thousand people is much, much easier than trying to work with 50 million.

This idea of expressing uncertainty using a probability distribution is a simple, but extremely powerful idea. Humanity owes a lot to the founding fathers of statistics, people such as Ronald Fischer, Jerzy Neyman, and Karl Pearson who came up with methods such as ANOVA and multiple linear regression and built up the mathematical foundations of much of modern statistics.

Without inferential statistics, most sciences would lack the tools to make informed decisions. Every time you hear about a study in medicine, psychology, economics, or otherwise, you should know this is possible due to inferential statistics.

How to Use Statistics

Okay, so now it's time to look at how you can use statistics. Once again, we'll be taking a high-level view and won't be looking at it from a statistician's or data scientist's perspective, but from a non-technical decision maker's perspective.

First of all, it's likely that you're already using descriptive statistics in some shape or form. Metrics like the mean and the median or plots and graphs like bar charts and histograms are something that most people can relate to. However, it's important to be aware of some of the pitfalls that could pop up, which we will discuss in greater detail a little later.

Then you have inferential statistics, which is useful in two situations. It can be used to answer particular questions about variable importance, but also to answer questions about differences in population.

By the end of the section, you should understand that statistics is always used to make a case supporting a hypothesis or story, and this involves the extensive use of descriptive statistics. While inferential statistics will often be employed in the analysis, it tends to be with descriptive statistics.

Thus, you as an executive, manager, founder, or decision maker need to be able to understand whether you can use statistics to answer a question and whether the story makes sense and to avoid all the pitfalls and fallacies that can be involved in statistical inference and the presentation of results.

Examples of Inferential Statistics

Let's start with a few examples of inferential statistics to give you a better idea of the concept itself but also to see how you can apply them to your business.

Hypothesis testing and statistical modeling are the two main things you will be using inferential statistics for.

Forecasting is another area and represents a set of tools and methods that allow for the prediction of future values. However, we won't be going into forecasting because it is a little more technical and advanced, and it can often be accomplished using machine learning.

Hypothesis Testing

Hypothesis testing is generally used for two things, namely, finding out the value of a population parameter or making comparisons. Determining the value of a population essentially means finding out something like the average height of a population or what their average income is.

Comparisons, on the other hand, involve exploring differences and/or similarities among similar parameters. For example, you could explore differences based on gender, comparing whether men earn more money than women in the same job.

Hypothesis testing is an extremely important tool for science. Every time, for example, you hear about clinical trials for new drugs, a hypothesis test is being employed to detect whether any significant effect is observed. When conducting hypothesis testing, we are essentially comparing two hypotheses: the *null hypothesis* and the *alternative hypothesis*. So, in the context of a clinical trial, the null hypothesis would be that a new drug has no effect on a disease.

The hypothesis test produces something called a p-value. This is the probability of the null hypothesis being true, given the data. If the p-value is low (usually we use a 5% threshold), then we reject the null hypothesis and accept the alternative hypothesis.

In business you mainly encounter hypothesis testing in the context of A/B testing. So, when someone is testing two headlines against each other to see which generates the highest click-through rate, for example, they are conducting a comparison-based hypothesis test.

See Table 6-1 for examples of statistical tests.

Table 6-1. Some examples of statistical tests.

Name	Use
T-test	Used to test the equality of the means between two groups
ANOVA	Used to test the equality of the means of multiple groups
Test of proportions	Used to test the equality of proportions in two groups
Wilcoxon signed rank test	Non-parametric alternative to the t-test
Smirnov-Kolmogorov test	Used to test whether a variable is normally distributed
Friedman test	Non-parametric alternative to ANOVA

Statistical Modeling

Statistical modeling is used mainly when you are interested to determine whether one variable influences another.

For example, let's say that you've conducted the hypothesis test regarding men earning more money than women in the same position and have come to the conclusion that the statement is accurate. However, you want to discover what other variables might affect income. To achieve this, you turn to statistical modeling.

Figure 6-2 shows an example of linear regression, the most famous statistical model.

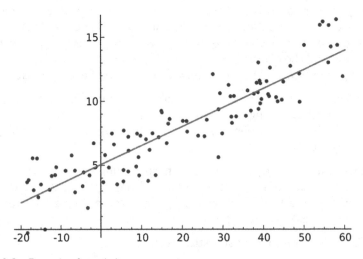

Figure 6-2. Example of simple linear regression

So, you enter these variables into the model and discover that income is also affected by age, experience, previous employment, and more. You can also find out to what degree these variables affect the response variable—or the variable of interest—which is income level in our case.

Have you ever heard of *linear regression, logistic regression,* or the *generalized linear model?* All these are examples of statistical models. These models form the basis of models used in other more specialized fields such as econometrics and psychometrics. See Table 6-2.

Table 6-2. Examples of statistical models

Name	Use
Multiple linear regression	Most common statistical model, used when the response variable is numerical
Poisson regression	Used when the target variable consists of positive integers
Logistic regression	Used for binary outcome variables, e.g., understanding the differences between two types of customers
Survival modeling	Used when we care about the time something will take until it fails, e.g., a mechanical part
Log-linear model	Model used to examine the relationship between more than two categorical variables

Misleading with Statistics

People tend to trust statistics blindly and this can be a problem. It's natural to be accepting of what seems like complete facts, especially when they are supported with lots of figures and charts and so on. However, the fact is that statistics can be used to deceive people. They can also lead to skewed results because of mistakes.

This section is essential because statistics is used in the wrong way quite often. We'll look at cases where statistics are being abused, and it's important for you to become aware of the many ways things can go wrong and also of the cases when people are trying to deceive you. This way, you'll be in a position to make better informed decisions.

Lying with Charts

So, let's start with one my favorite ways to mislead with statistics which is to blatantly lie using charts that contain spurious correlations and seem to make sense on the surface, but in reality, they don't make any sense at all.

Take a look at Figures 6-3 and 6-4. Figure 6-3 shows a correlation between the number of people who drowned by falling into a pool and the number of films Nicolas Cage appeared in. Figure 6-4 correlates US spending on science, space, and technology with suicides by hanging, strangulation, and suffocation.

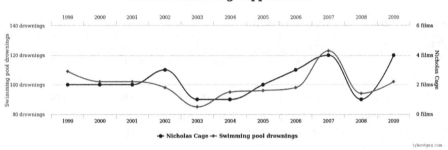

Figure 6-3. Nicolas Cage seems to be driving swimming pool drownings worldwide[1]

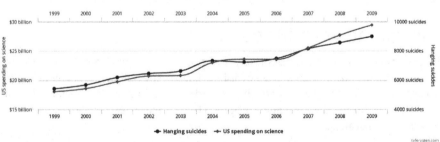

Figure 6-4. US spending on science, space, and technology correlates with suicides by hanging, strangulation, and suffocation[2]

---

[1] Graph by Tyler Vigen (http://tylervigen.com/view_correlation?id=359) is licensed under CC BY 4.0 (https://creativecommons.org/licenses/by/4.0/).
[2] Graph by Tyler Vigen (http://tylervigen.com/view_correlation?id=1597) is licensed under CC BY 4.0 (https://creativecommons.org/licenses/by/4.0/).

Now, at first glance, these might look like interesting statistics. The first chart seems to imply that more people drowned when Nicolas Cage starred in more films. Based on this chart alone and without the understanding that statistics can be misleading, you might conclude that people may have actually killed themselves to get away from Nicolas Cage movies.

The second chart implies that US spending on science, space, and technology has increased as the number of suicides by hanging, strangulation, and suffocation increased. So, you could draw the conclusion that investing in R&D is a sure way to increase deaths in the overall population!

When you write out the conclusions, it's easier to realize that they really don't make a lot of sense. And yes, these are extreme examples to show how statistics can be used to mislead, but many similar examples exist where the correlations are spurious, even if it's not that obvious, which is why you need to be careful.

So, what does it mean that the correlations are spurious? It means that two variables are correlated but the underlying theory as to what's happening doesn't exist. They are simply two variables with no real connection that have been correlated. And let's not forget that correlation does not imply causation.

In other words, the number of Nicolas Cage movies has absolutely no connection with pool drownings in the real world, making it a spurious correlation. Conversely, if we were to compare Nicolas Cage's earnings with the number of movies he starred in, then the correlation would make sense because there is a connection between the two.

The second chart goes even further in its attempt to trick people because the X-axis has been reversed. In other words, the numbers decrease along the axis instead of increasing, despite the fact that the Y-axis has values that increase as you go higher. This is an attempt to further force a correlation.

So, statistics is a great tool, but it isn't a magic wand that can provide answers no matter what. Using statistics also doesn't mean that you can get away with not having the right research design. You also need to have the right framework when looking into problems.

If you want to see more examples of spurious correlations, visit http://tylervigen.com/. It's actually quite funny because in all these correlations, the mistake is always the same, namely, that the theory behind the two variables being studied is not there.

Remember, just because there seems to be a high correlation between the two variables, it doesn't actually mean that the two are related.

The problem is that while all these examples have spurious correlations, there are many situations where it isn't quite as obvious.

For example, let's say that there's a high correlation between the amount of time someone has lived with diabetes and the number of people who die before they turn 60. On the surface, without digging deep, you might conclude that the longer someone has had diabetes, the more likely they are to die before they turn 60. It makes for a pretty chart and everything.

However, if you start digging deeper, you discover that the deaths include a wide range of causes, including car crashes, murders, suicides, and so on. In other words, part of the second variable is completely unrelated, even if part is. The result is the same though, namely, that the decision maker studying a chart like that will come to the wrong conclusion.

Another way to lie with charts is to use different kinds of scales. In the example in Figure 6-5, we have two charts with two bars each, representing type A and type B. Both charts use log scales without explicitly stating they are doing so. Now, by using different scales, you can make two items appear closer to each other than they actually are.

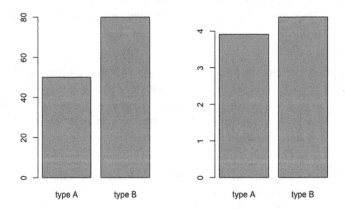

Figure 6-5. Example of how a logarithmic scale can skew perception of the differences between groups

Another easy way to lie with charts is to play around with the scale. For example, in Figure 6-6 you can see how the image on the left shows a rising trend and the image on the right shows a more or less stable trend. The data is the same in both cases. The only difference is the degree of zoom.

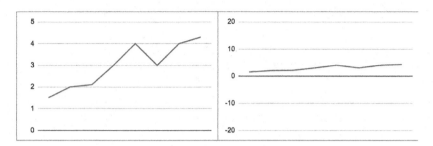

Figure 6-6. Same data, different view

Misleading Using Descriptive Statistics

Now, let's take a look at how you can mislead someone using descriptive statistics. One way to easily lie is by using measures of centrality and summary metrics. There are, in essence, three main measures of centrality, namely, the mean, the median, and the mode.

The mean is the average of a set of numbers, namely, where you add all the numbers together and divide by how many numbers there are. The median is the middle value in a list of numbers, while the mode is the value that occurs most frequently in a list.

Now, in a symmetrical distribution, the mean, median, and mode fall on the same point. However, when we have a positive skew, the mean is lighter than the median, and the median is lighter than the mode, and the opposite happens in a negatively skewed distributions. See Figure 6-7.

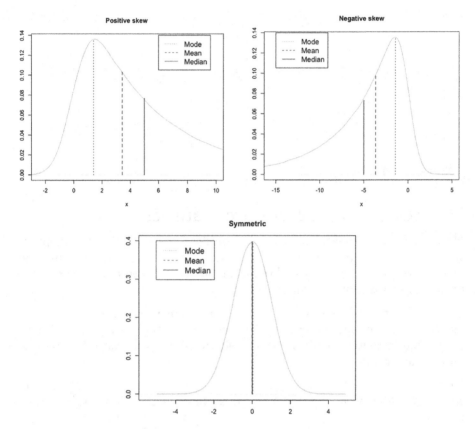

Figure 6-7. Skewness and the relationship between the mean, the median, and the mode

This is important because many distributions in real life are not symmetrical. For example, let's say there is a company where the CEO earns $300K per year, while the rest of the employees—a total of ten—earn $50K per year.

Depending on the image you want to project, you can present different figures. For example, in our situation, the median is $50K per year, but the mean is $72K per year.

A malicious company might advertise the mean salary as being $72K per year when looking to hire someone, but in reality, they aren't willing to pay employees other than the CEO anything over $50K. But this way, they could trick somebody into thinking they could eventually make $72K per year, and they wouldn't even have to come out and say it like that as the figures imply it.

Biases in Sampling

Now we're going to look at biases in sampling. There are a number of sampling biases, but we are going to focus on five, namely, selection bias, area bias, self-selection bias, leading question bias, and social desirability bias. These are the most common in a business setting.

Selection Bias

Selection bias is the most important, and the other biases can be considered related or even seen as subsets of this type of bias.

This type of bias exists when the sample is not representative because there is some kind of process taking place that makes sure the subjects or the items in the sample are special. Many other subjects or items are not included.

An interesting story about selection bias is the one of Abraham Wald, who was a statistician. During World War II, the government reached out to him to ask him to improve the survival rate of US aircraft. Wald was smart, and the first thing he did was to look at the prior analyses that had been conducted.[3] And he noticed that selection bias was taking place.

The previous investigators had been analyzing aircraft that survived and recommended that more armor should be added to the areas that had received the most damage, thereby increasing their protection. So, where the planes were most shot and torn up, new armor was mounted to protect those areas.

The result was that even more planes went down. The additional armor made the planes heavier, leading to them being less agile and slower. And the planes still returned with damage in the same areas.

The problem was that the investigators had based their conclusions on planes that had been damaged in areas that were clearly resilient. After all, they had taken damage in those areas and still returned to base.

Wald realized that what needed to be protected was the areas that had not taken damage and were sensitive. Once armor was added to those areas, survivability increased.

It's a very interesting example of using statistics in a highly intelligent way without even running an analysis. All he had to do was realize that this selection bias was in play and then he fixed it.

---

[3] Marc Mangel and Francisco J. Samaniego, "Abraham Wald's Work on Aircraft Survivability," *Journal of the American Statistical Association* 79, no. 386 (1984): 259-267, https://people. ucsc.edu/~msmangel/Wald.pdf

Area Bias

Area bias occurs when you try to generalize results from one area to another area, and this is especially problematic when attempting to cross cultures.

So, for example, you conduct marketing research in Europe and you automatically assume the same results will hold for the United States. Unfortunately, this doesn't take into account that a wide range of factors, including cultural and financial, could affect the validity of the results.

Self-Selection Bias

In self-selection bias, a person's decision to take part in a study could be correlated with traits they already have and which could affect the study, making the participants a non-representative sample.

This might sound complicated, but it's actually a lot simpler than it sounds. For example, if you were to set up a booth to ask people about their grooming habits, the people who respond are more likely to be people who like taking care of themselves and spend more time primping in the mornings.

The problem is that you need to talk to both people who primp and also those who are less interested in grooming in the mornings for the sample to be representatives.

Leading Question Bias

Leading question bias is about how questions are phrased and asked, but also what the answer options are.

So, if you were to ask

Don't you think high school teachers are paid too little?

A) Yes, they should earn more.

B) No, they should not earn more.

C) No opinion.

Most people would answer A because of the way the question was phrased. Thus, with leading questions, you end up with biased results.

Social Desirability Bias

Social desirability bias occurs when people want to seem better than they are. For example, if you ask people in survey about how often they shower or how frequently they recycle, your data will be biased because no one wants to admit to doing something that is socially undesirable.

In other words, even if someone showers every 2 weeks, they're going to tell you they do it once a day or every 3 days or whatever they think is a more socially acceptable number. Whatever it is, it's rarely going to be the truth.

Sampling Bias in the Real World

The previous examples were all quite simple. However, the polling for the 2016 American Presidential Election is a good case where these sampling biases combined to affect the results. Another example is Brexit.[4]

In both cases, the pollsters' results weren't that accurate. Most of the explanations of what happened revolved around biasing in sampling. There are all sorts of theories regarding what happened.

For example, the Pew Research Center stated the following:

> One likely culprit is what pollsters refer to as **nonresponse bias**. This occurs when certain kinds of people systematically do not respond to surveys despite equal opportunity outreach to all parts of the electorate. We know that some groups—including the less educated voters who were a key demographic for Trump on Election Day—are **consistently hard for pollsters to reach**. It is possible that the frustration and anti-institutional feelings that drove the Trump campaign may also have aligned with an **unwillingness to respond to polls**.[5]

Another theory was that there was self-selection in online polling, saying that people who tend to answer online polls were more likely to support a certain candidate over another. There was also the assumption that people who supported the other party wanted to avoid polls altogether.

It once again shows how statistics can also go wrong not because of the math or the models but due to factors outside of our control, namely, sampling biases.

This is similar to spurious correlations in a way, because statistics is an applied science and doesn't happen in a vacuum. If you don't have the right data, if you don't have the right theory, you will still come up with a set of numbers but they won't mean anything.

---

[4] I have also written about this on my blog: http://thedatascientist.com/election-forecasting/

[5] Andrew Mercer, Claudia Deane, and Kyley McGeeney, "Why 2016 Election Polls Missed Their Mark," *Pew Research Center*, November 9 2016, www.pewresearch.org/fact-tank/2016/11/09/why-2016-election-polls-missed-their-mark/

Lying with Inferential Statistics

So, while it might not be that easy to lie with inferential statistics, those of you who might have harbored the illusion that it's not possible at all will be disappointed because it can be done.

It happens in large part in research. When you test a hypothesis, you get a p-value, which you want to be below a certain threshold for the test to have statistical significance. Usually this threshold is 0.05.

There's a theory about the cult of the p-value[6] that says there are a very large number of research papers out there with p-values suspiciously close to 0.05. Essentially, they're just below 0.05 so they can meet the condition of statistical significance. This might make the results significant, but only marginally so.

The theory is that this is being done on purpose. In academia, the environment is very competitive and people have to publish or they'll lose their job. Thus, many people are under pressure to make their p-values look more significant than they might actually be.

One way to do this is to run multiple experiments until you get the results you want, and another way is to massage the data. In some cases, though, you might find research where the results are significant but the researcher's intent is not to deceive, but it could be that the model assumptions are not being met.

All models and statistics have some assumptions, and unfortunately there are many research papers out there where these assumptions are not being properly accounted for.

IN DEPTH: BAYESIAN VS. FREQUENTIST STATISTICS

Maybe you have heard of the term "*Bayesian.*" There are many models that include the word "Bayes": "Naïve Bayes," "Bayesian networks," "Bayesian inference," and so on.

All these refer to a school of thought behind statistics. It all started with the British mathematician *Reverend Thomas Bayes*, who came up with this simple formula:

$$P(A|B) = \frac{P(B|A)P(A)}{P(B)}$$

---

[6] www.press.umich.edu/186351/cult_of_statistical_significance

This formula describes the probability of event A taking place, given that B has taken place as well. This was later developed further by Pierre-Simon Laplace and Sir Harold Jeffreys into what is now forming the foundations of Bayesian statistics.

One of the main ideas behind Bayesian statistics is that it is possible to incorporate *prior* knowledge about something into our analysis, be it a model or a significance test. On the other hand, *frequentist statistics* (which is the most popular statistical theory) do not accommodate that, neither do they allow it.

So, for example, let's say that you want to run some experiment to prove that ghosts exist. You have two hypotheses (ghosts exist or not). A Bayesian statistician can pre-assign a very small probability to the hypothesis that ghosts do not exist based on common sense. Hence, the experiment should find overwhelming evidence in favor of the ghost hypothesis before the test is significant. A frequentist doesn't have this choice.

Let's see another example. Frequentist statistics are formulated under the concept of an infinite number of experiments. For example, if you flip a fair coin a large number of times (thousands or more), then the actual probability of flipping heads will converge to 0.5.

However, how can this analogy work for events such as football games or elections? These events are unique. A game between two teams in a specific point in time is not going to be repeated. Even if it were, the parameters underlying it (e.g., how tired the players are or the members of a political party) might change. We can't perform an infinite number of these events. A frequentist is in trouble in this case, but this is not a problem for a Bayesian. In Bayesian statistics, probabilities express *subjective beliefs* of events happening, not repeated experiments.

Bayesian statistics have proven to be very useful and to also provide a theoretical framework for parts of machine learning. There are still many disagreements in the statistics community as to which theory is the best. In practice, frequentist statistics are used more often, but there are also many applications based on Bayesian statistics. Once again, it's about finding the right tool for the job!

A Short Introduction to Machine Learning

Machine learning has been defined in many ways by different researchers. The following are definitions from some of the most prominent researchers in the field:

> Machine learning is programming computers to optimize a performance criterion using example data or past experience.

> —Ethem Alpaydin[1]

---

[1] Ethem Alpaydin, *Introduction to Machine Learning*, 4th ed. (Cambridge, MA: MIT Press, 2014).

© Stylianos Kampakis 2020
S. Kampakis, *The Decision Maker's Handbook to Data Science*,
https://doi.org/10.1007/978-1-4842-5494-3_7

> *The goal of machine learning is to develop methods that can automatically detect patterns in data, and then to use the uncovered patterns to predict future data or other outcomes of interest.*
>
> —Kevin P. Murphy[2]

> *The field of pattern recognition is concerned with the automatic discovery of regularities in data through the use of computer algorithms and with the use of these regularities to take actions.*
>
> —Christopher M. Bishop[3]

So, while these definitions are slightly different, but in one way or another, they involve algorithms that can learn from experience and data by identifying patterns or using other methods. They then use what they have learned to take action.

Let's take a look at a simple example to gain a better understanding of how machine learning works. In Figure 7-1, I'm showing you a fictional dataset that consists of images of cats and dogs.

Figure 7-1. Classification example.

---

[2] Kevin P. Murphy, *Machine Learning: A Probabilistic Perspective* (Cambridge, MA: MIT Press, 2012).
[3] Christopher M. Bishop, *Pattern Recognition and Machine Learning* (New York: Springer, 2013)

If I ask you which images depict dogs and which ones cats, it is ridiculously easy to do. But how does it happen?

First of all, your brain has been exposed to a huge number of images of cats and dogs, but also examples of those animals in real life. You have trained yourself to be able to discriminate between these two and other animals through countless rounds of feedback from sources like your family, school, and TV.

However, this whole process isn't quite as simple as it seems. Your brain has been optimized through thousands and thousands of years of evolution to make these associations easily. Even if you knew nothing about cats and dogs, your brain would quickly figure out that these two entities look like different animals. Your brain can aggregate these images very quickly and it immediately understands that while they are different, they all refer to the same species.

For a machine, though, this is much more difficult, which is why machine learning is so important because it's about imitating this process and replicating it within a machine.

The Main Advantage of Machine Learning

Learning from data is very easy for the human brain to do but difficult for a machine to do because it requires the right algorithms and lots of data to achieve.

However, what is easy for a machine is implementation and automation. Once you've got the algorithm, it's easy for a machine to go through thousands or even millions of images without experiencing any fatigue, which is impossible for a human brain to do.

It means you can create something very powerful with superhuman capabilities, even if we're talking about a narrow field of application, such as image classification. And this is exactly what makes machine learning and big data such a powerful combination.

Types of Machine Learning

Now, we're going to take a quick look at the different fields in machine learning (Table 7-1). The main fields in machine learning are **supervised** and **unsupervised learning**. Reinforcement learning has been gaining ground and related fields such as recommender systems can also be linked to machine learning, but they are more niche and not within the scope of this book.

Table 7-1. Types of machine learning

Name	Use
Supervised learning	Use a model M to predict target variable Y from input X.
Unsupervised learning	Find patterns in variable X (clustering is the most common example).
Reinforcement learning	Teach an agent how to optimize a reward function over the environment. For example, a video game AI can take 10 different actions, which need to be done as a response to the environment in order to maximize to reach a goal (e.g., maximize a score or beat a human player).
Active learning	Supervised learning with active feedback from a human.
Semi-supervised learning	Supervised learning, where some examples are unlabeled. For example, a dataset of 10000 images, where 1000 do not have a label, can be analyzed through semi-supervised learning techniques.
Recommender systems	Find a set of items to offer to a user that the user is likely to like (e.g., movies to watch, products to buy, songs to listen). Heavily related to supervised learning.

Supervised Learning

Supervised learning can be classified into classification and regression, which are both quite similar. In supervised learning, there is some form of supervision of the algorithm. We provide the algorithm with a raw dataset, but then we also provide it with a target. The latter can be a single variable or multiple variables.

Then, the algorithm attempts to understand how to match the input to the output. For example, you provide the algorithm with customer data and then it tries to determine if someone will make a purchase or not.

The supervisor oversees what the algorithm is doing. Every time the algorithm makes a mistake, the supervisor tries to correct it. It's basically like the algorithm is learning under the guidance of a teacher.

Unsupervised Learning

Unsupervised learning, in the context of machine learning, usually refers to clustering. In this case, we provide the algorithm with raw data, but instead of providing a target, the algorithm is allowed to do its own thing essentially and identify patterns in the data provided. It tries to find any regularities that seem significant, but there is no supervision involved.

This type of learning is a very difficult problem because it's ill-defined and almost always requires a human to interpret usually toward the final stage of the pipeline. For example, the algorithm might identify patterns that are irrelevant to what you are trying to do, which is why you need to drill down into the results and try to define them a little better.

Supervised learning, on the other hand, is a bit easier because you can run an algorithm and you can immediately determine how it works. You can analyze customer data, for example, and predict retention rates with an 80% accuracy. This is what makes supervised learning one of the most popular fields in the machine learning family.

It's important to note that some the terms we are using such as classification, regression, and clustering are used in statistics as well as machine learning. However, even if the terminology refers to almost the same thing, there are still some differences.

In statistics, interpretability is important, whereas it is of little consequence in machine learning. Likewise, in machine learning, we are interested in the predictive performance of the algorithm, while in statistics this is important.

The terminology might seem a little confusing when comparing the terms from statistics to machine learning because there are some models like logistics regression models that are classification models and are treated as classification models within the context of machine learning. The differences are mainly the result of different histories to the fields, so don't be confused if you hear people with different backgrounds using these terms in slightly different ways.

A Closer Look at Supervised Learning

Supervised learning is broken down into **classification** and **regression**. The difference between the two is quite simple. In classification we try to identify if something belongs in a specific category, while in regression we try to predict a real number.

An easy example of a classification model is one where we attempt to determine whether an email is spam or not. It's a task that's found in every introductory machine learning textbook. Another popular task is image recognition or object identification, such as "is this an image of an animal?" or "does this image contain a human?" or "where in this picture can you find human faces?".

Classification can also be about other things. Determining whether a stock will rise or fall is one such application, and it's usually referred to as binary classification because there are two categories—rising or falling. Another application is to determine whether someone will click on an ad or will the temperature rise above 0 degrees.

You can have multiple categories too in some problems. For example, if you want to predict elections, in most cases, you have more than two election parties. In that case, you might have four or five classes, with each class representing one of the political parties running in the election.

Regression, on the other hand, attempts to predict a real number. So, instead of asking whether the price of a stock will rise or fall, we attempt to determine what the price of the stock will be in 2 weeks. Or, if we are attempting to determine customer demand for a particular product, we could attempt to figure out what demand will be like in 5 days, for example. Maybe instead of attempting to understand if someone will click on an ad, we want to find out how many people will click on an ad—it might be a related problem but it's still a different problem.

Likewise with temperature. A vague idea of whether it will rise or not might not be sufficient for your needs, so you might attempt to determine what the exact temperature will be in 3 weeks.

These two types of problems—classification and regression—can be found in many fields and in many different contexts.

It's important to note that the same problem can be tackled using different approaches. Let's say that we work for a hedge fund and we want to predict the price of stocks so we can make a nice profit.

This problem can be tackled using statistical forecasting, machine learning regression, or even classification with the latter attempting to determine the direction price will head in, that is, either up or down.

The performance ceiling—or the best performance you can get—for each approach, though, will be different. This can have different effects on your strategy, depending on what you want to achieve.

For example, you might first attempt to use a classification model to predict the price of stock. But you realize then that you'll only know whether the price will rise or fall. This doesn't lend itself well to straight trading, so you might consider spread betting as an alternative. However, you might decide to take the slightly more difficult approach of a regression model because the increased profits will make it worth your while.

The key is to understand that a problem is often not just a single problem but a series of other problems within your main problem. Think of it a bit like a LEGO model. Each model is a whole—a castle, a helicopter, a spaceship—but it's built from separate blocks. In this case, the problem is the castle made up of sub-problems or blocks.

It also pays off to try multiple approaches to determine which one is the most effective for your needs. Again, it's like one of those multi-model LEGO sets. You can use the same bricks in different ways to achieve the outcome you need.

Except that, unlike with a LEGO set, you don't have instructions or even inspirational pictures to help you on your journey—you have to keep testing different assembly approaches until you end up with what you need.

However, if you have a creative data scientist around to help you, it'll be like having those LEGO instructions and pictures on hand, making life much easier.

It's also important to choose your battles wisely. If you find that a regression model will generate better performance, even if it's more difficult to do and wasn't your initial choice, it's still a good idea to switch to ensure better results.

A Better Understanding of Unsupervised Learning

Unsupervised learning is about learning patterns from data without any kind of guidance. The main application is clustering, which is an automated way of creating groups, like with the images in Figure 7-1.

So, you can feed a bunch of animal pictures into an unsupervised learning algorithm, and it could group them automatically into two categories (cats and dogs) without any additional feedback.

One of the most prominent uses for unsupervised learning algorithms in business is for customer segmentation. So, you feed the algorithm customer data, and it identifies patterns and segments your customer base accordingly.

Another application is anomaly detection, which some consider to be its own separate category because it doesn't use just unsupervised learning but also borrows some techniques from supervised learning. However, I've included it here because it's good to know that it exists.

Dimensionality reduction is another application which involves reducing a large number of variables into what are called "factors" or super variables. So, you can take 20 variables, for example, and group them under one factor. It's a little like clustering, but for variables. The main reason to do this is because it makes it easier to understand the data.

Examples of Clustering

As already stated, unsupervised learning and clustering are fundamentally very difficult problems, mainly because they are ill-defined. So, let's say we are given the objects in Figure 7-2 to cluster.

Figure 7-2. Clustering example

The problem is that there are numerous approaches. You could cluster according to color. Or, you could cluster by type, namely, based on whether the object is a fruit or vegetable. Or maybe you want to cluster based on shape.

However, if you want to cluster, for example, based on color, this doesn't mean that the other approaches are wrong. All the patterns are valid. The question is whether they are useful, and the answer depends on what your goal is.

Thus, clustering algorithms attempt to identify these patterns, but sometimes the results aren't that useful or aren't relevant to what you need. It's why you always need a bit of experimentation before you can get it right. At the end of the day, though, if you know what you're doing and you try a few different algorithms, you will eventually find one that works.

Unsupervised learning has been around for decades so there are a wide range of very powerful algorithms that can identify patterns in data if they exist. However, it's important to keep in mind that just because you think objects should be clustered in a certain way doesn't mean the first algorithm you try will come up with the same result.

Examples of Anomaly Detection

Anomaly detection is about anomalous or unusual behavior in different contexts. Figure 7-3 is a time series, and by running an anomaly detection algorithm, we can see that the search volume for the term 'puppy' spikes every January.

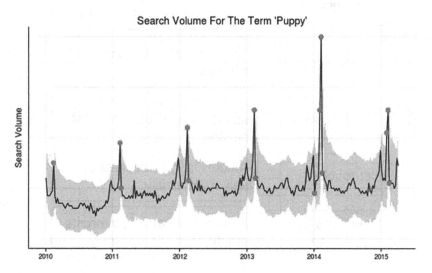

Figure 7-3. Anomaly detection example over a time series[4]

How is this useful? Well, if you own a pet shop or if you run a dog shelter, you'll know when is the best time to match puppies with new families or to increase your puppy-related inventory.

Likewise, this kind of analysis is used for a wide range of applications. It can be used in retail for a wide range of products, or it can be used to analyze social media activity to understand whether people talk about particular subjects and if a topic has created a spike.

Anomaly detection is extensively used in cyber-security to detect hackers. So, if 99% of the traffic going to a server is normal and then you have 1% that is exhibiting anomalous behavior, anomaly detection algorithms will flag the suspicious behavior so action can be taken accordingly.

---

[4] Figure is licensed by John Greer (https://jgeer.com/anomaly-detection-how-to-analyze-your-predictable-data/) under CC BY 3.0 Unported (https://creativecommons.org/licenses/by/3.0/)

Examples of Dimensionality Reduction

In Figure 7-4, you can see a summary of the Big Five personality traits. It is one of the most successful theories in psychology.

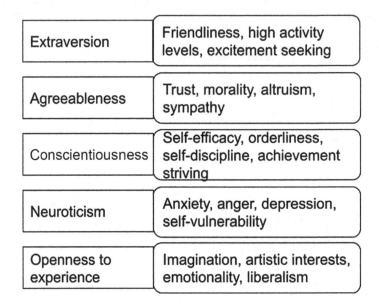

Figure 7-4. Summary of the Big Five personality traits

So, what does the Big Five personality theory have to do with dimensionality reduction? Well, this theory was derived by running factor analysis, one of the most prominent dimensionality reduction techniques.

This personality theory was the result of people answering a large number of questions in a questionnaire and then applying factor analysis to group these questions into five overall factors. As you can see, according to this theory, personality is broken down into

- Extraversion
- Agreeableness
- Conscientiousness
- Neuroticism
- Openness to experience

This is an excellent example of where dimensionality reduction can be used. It is also sometimes used in supervised learning because it can de-noise data, but this is more of a technical issue. However, it can also be used in any other context where you have a large number of variables.

For example, in my PhD thesis, I worked with GPS data from the training sessions of the Tottenham Hotspur football club. The GPS units we used collected over 100 different variables, which the sports science team found very difficult to interpret. So, I resorted to supervised principal component analysis and reduced them down to five or six factors that correlated with injury,[5] which made the results much easier to understand.

As you can see, it can be applied to many things. There's nothing that says you can't do the same with user demographics or customer data and so on.

INTERPRETABLE MACHINE LEARNING

Machine learning has often been called a black box, and rightly so. Not only the algorithms can seem difficult to understand to the uninitiated, but also the algorithms are usually designed for performance, and predictive power, but not interpretability.

However, there is a new niche in machine learning that is set to change all that: interpretable machine learning.

Interpretable machine learning is focused on coming up with techniques and methods in order to make machine learning models more transparent. For example, one such simple technique to use is called the "surrogate model" technique. A decision tree is fit on the predictions of a more complicated model (e.g., a deep neural network). The decision tree can be easily interpreted, even by someone who is not a data scientist, and can very quickly help you spot out the most important variables and how they interact.

Two other more advanced techniques are LIME and Shapley value explanations. LIME stands for Local Interpretable Model-Agnostic Explanations and is a technique that can be used to explain any outcome. Shapley value explanation is a technique inspired by game theory, which can be used to understand the importance of variables and how they interact within a model for it to reach a particular conclusion.

In Figure 7-5, you can see how, for example, LIME can spot the most important words for text classifier. The text classifier is trying to discriminate between posts that are talking about atheism and posts that are talking about Christianity. The algorithm can produce a list of importance of the different words, which are then highlighted in the text.

---

[5] http://thedatascientist.com/supervised-pca-practical-algorithm-datasets-lots-features/

Interpretable machine learning methods will become more and more important as algorithms are being used more and more often to make decisions in all parts of our lives. You can read more at https://thedatascientist.com/interpretable-machine-learning/

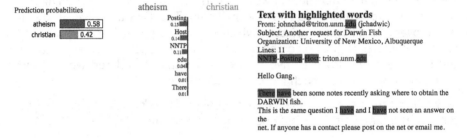

Figure 7-5. Example of how LIME can examine feature importance[6]

Problem Solving

In this chapter, we're going to look at solving a problem from the decision maker's point of view, as this is what this book is about. So, we aren't going to look at how data scientists solve problems but how you will solve the problem working alongside a data scientist.

No one expects the *domain expert*—a.k.a. you—to be qualified in using data science. However, you need to be able to think *like a data scientist* when solving a problem, because it will help you improve how you define the problem, as well as making it easier for you to find the right people and to manage them.

You need to be able to pose a problem and, as previously mentioned, one highly effective approach is to turn the problem into a question. You also need to understand whether there's any value in solving the problem. Everything takes time and money, so you need to understand whether the results of the problem will be applicable to your business. Lastly, you need to understand who you should hire to help you and if you need more than one person.

Understanding Whether a Problem Can Be Solved

To figure out if something can be solved, you first need to be aware of everything that can go wrong. From experience, I've discovered that businesses face two main issues when they want to use data science.

© Stylianos Kampakis 2020
S. Kampakis, *The Decision Maker's Handbook to Data Science*,
https://doi.org/10.1007/978-1-4842-5494-3_8

First, they often have the wrong expectations, so the actual solution can turn out to be either much easier or much more difficult than they think. Secondly, they don't have the right data.

A data scientist can help you understand whether the problem can be solved at all, whether the problem is difficult or easy to solve, and how much time and resources will be required.

So, what you need to do is to find the right person—preferably someone you trust—and work alongside them.

What you should never ever do is make assumptions without consulting a data scientist! Many issues arise because people make erroneous assumptions, especially around technical matters, such as the complexity of the project or the timeframes involved. It's better to ask if you are uncertain about something—and even when you're not—than to make assumptions.

Quick Heuristics

We're now going to look at a few heuristics to help you determine if data science can be applied to a specific problem.

First of all, you need to determine if the problem can be phrased as

- A statistical modeling problem

- A hypothesis test

- A supervised learning problem

- An unsupervised learning problem

Let's take a look at some examples.

Statistical Modeling Problem

Statistical modeling should be used when you have a situation in which you are trying to determine whether variable X is important for variable Y and what the relationship between the two is. Statistical models can tell you which factors are significant and what the direction of effect is.

So, for example, you might have noticed that older people seem to like your product more. Thus, you might want to determine if age plays a role in the purchasing decision. In this case, you'll want to hire a statistician since you will be using statistical modeling.

As it was mentioned before, the focus of statistics is on transparency. When you care about understanding the relationships between variables, you are probably going for statistics.

Hypothesis Testing

Hypothesis testing is the approach you want to turn to when you want to compare two groups. This is if you want to figure out if X and Y differ. An example of this is A/B testing.

Let's say you want to roll out a new feature for your product and you don't know if it's a good idea. All you need to do is create two different groups and compare them with each other.

Likewise, you might want to compare two existing groups. For example, you might want to study gender, regardless of context. You might want to see if gender has an impact on sales, so you would create two groups, one of each gender, and make the comparison.

Hypothesis testing is closely related to statistical modeling, and sometimes the same problems can be answered through different hypothesis tests or models. In both this case and the previous one, you know you need a data scientist with skills in statistics.

Supervised Learning

This approach is a little more difficult. We've already mentioned two scenarios, namely, classification and regression, so these are clearly situations where you can use supervised learning. You have a dataset, and you want to use some input variables to be able to predict an outcome variable Y.

A good heuristic is that you can use supervised learning in any scenario where you want to automate something a human is doing. If you have someone sitting there detecting spam or labeling images and you have a dataset, you can feed that data to a machine along with examples from the human, and the machine will attempt to mimic the decision-making process of the human.

Also, supervised learning is a great option when you want to predict future values, that is, predictive analytics—predicting someone's risk of bankruptcy in the future (i.e., their credit score), demand for products in the future, or the price of a stock.

Note that supervised learning algorithms are usually not very good in explaining how they reached a decision. Machine learning works well, but it is mainly a black box. Hence, if understanding your data is of paramount importance, you might have to perform exploratory data analysis or create some statistical models as well, such as linear or logistic regression.

Unsupervised Learning

Unsupervised learning can be useful in quite a few scenarios. First, the obvious one is that unsupervised learning is useful when you know there are groups in your data. For example, customer segmentation is such a scenario.

Dimensionality reduction is another use case. If you have a very complicated dataset with a lot of variables and you aren't sure what to do with it, then turning to dimensionality reduction techniques, such as factor analysis, is a good idea.

Besides the fact that unsupervised learning can help you understand the data better, the outputs of unsupervised learning can also be used as inputs to supervised models to improve performance.

A Few More Heuristics

So, if you have images and audio that need to be classified, then you will turn to deep learning. The latter has been commoditized to a significant degree and tech giants like Google and IBM offer deep learning as a service.

If you have a B2C business, you will have to look into recommender systems and market-basket analysis. Recommender systems have become the norm for consumers.

Market-basket analysis (or association rule mining) represents a set of algo-rithms that try to identify frequent patterns in order to find associations between the features of a user and their behavior. These are patterns of the form; if someone buys X, then they are likely to buy Y.

If you want to develop a chatbot, there are platforms out there to help you do that, such as Dialogflow.[1] It should be noted that while chatbots have come a long way, they aren't as advanced as many people claim them to be. Some intelligence and machine learning go into them, but they still have a long way to go. However, they aren't difficult to build.

Finally, sometimes you want to forecast future values of a time series. Forecasting time series is a special field on its own. Statistics has a subfield that deals with these types of problems, but people also turn to machine learning to answer these types of questions. So, in practice, you might need a blend of two approaches. Because forecasting is very challenging and there are many ways to make mistake, I usually advise people to try and see whether the same problem can be treated as classification or regression.

---

[1] https://dialogflow.com/

When Heuristics Fail

The aforementioned heuristics don't work every time. However, when they don't work, it's usually because of one of the following.

A Vague Project Plan

Heuristics will fail if your project plan is vague or ridiculously grandiose. You hear people saying things like they want to build an AI capable of evaluating business plans with forecasts 10 years into the future and so on or that they want their AI to be able to predict everything.

Clearly, they've watched one too many episodes of *Star Trek* and have forgotten that technology isn't quite that advanced yet. It's certainly not a magic bullet capable of fixing any and every problem, and this holds true for machine learning and AI too.

So, it's much more effective to have a plan with a narrow focus, which also keeps things more manageable and realistic in terms of deliverables.

Developing Skynet to Kill a Fly

Another problem is when you try to develop an AI to get rid of a fly and it decides to use the entire world's nuclear arsenal to do so, also known as trying to use super advanced machine learning to do something a human can do easily.

For example, maybe you want to create a chatbot that understands conversation and books your appointments. It's a great idea, but you have to consider the time and resources it would take to develop and if the end result is something people would actually use. You can still go ahead, of course, but you need to be aware of the difficulties involved in developing this type of software.

Another fairly complicated idea is trying to automate data collection. It makes sense to try to automate data collection in different contexts, but in most cases, it's more effective to spend a little money to crowdsource it.

I've seen quite a few startups fail because though their idea clearly had value, the development costs in terms of time and money were too large for a newly founded company. And they were trying to solve problems that can be solved more easily by humans, even if it is slightly more expensive.

Lack of the Right Data

Another common problem is that businesses try to hire data scientists when they don't have the right data or the data is in very poor shape. Then the data scientists spend an inordinate amount of time trying to whip the data into shape instead of actually working toward a solution.

Other Considerations

There are a few other things you need to consider, namely, data quality, data volume in terms of number of variables, and data size.

Data can go wrong in many situations. You might not have a very small sample, for example, or maybe the quality is bad, or maybe you don't have the right variables.

Someone once asked me to examine the differences between gender and pay on their job platform. Unfortunately, gender was a variable they didn't collect so we only had first and last names to work with. The solution was to attempt to guess gender from someone's first name, but 30% of the names were unisex, so we had to scrap the data.

Their approach didn't make sense at all. If they knew they were interested in differences between genders, why not ask people their gender? It makes no sense at all and is downright ridiculous.

You definitely don't want to take the same approach. So, try to follow the principles we discussed in Chapter 2 and try to think about how you will be using your data before you start collecting it. *Failing to plan is planning to fail.* That's why it's good to have a data strategy from day 0.[2]

What Problem Do You Really Need to Solve?

Sometimes the problem you think you are solving is not the one you need to solve. Let's say you want to predict the price of Bitcoin to make a profit. This is a very difficult problem—I've done a lot of work on it so I should know.

However, if your goal is to profit from Bitcoin, this is actually a different problem. Maybe you're interested in whether the price is climbing or falling. Or maybe you want to use reinforcement learning to create an algorithm trading bot to do it for you.

The important bit here is that the actual problem you are trying to solve will influence the final success of your model! Good data scientists know how to pick their battles wisely, so as to maximize success. We saw an example earlier where the data acquisition considerations interact with the business plan. The same holds here.

Always make sure to discuss these issues with the data scientist. I've seen months being wasted, because the client said they needed to do one thing (e.g., regression), but in reality the problem would have been solved more easily if the data scientist had tried to solve a different problem (e.g., turn the regression problem into a three-class classification problem).

---

[2] http://thedatascientist.com/data-strategy-startups/

EXAMPLE: THERE ARE MANY WAYS TO REACH THE TOP

A proverb says "there are many ways to the top of the mountain, but the view from the top is always the same." This couldn't be truer than for the case of data science. As it was explained in the chapter, sometimes a problem is not a problem. It's not uncommon for a business to confuse what they want with what they need.

I had a client once who asked me if I knew how to code a recurrent neural network. I answered yes, but I was curious to see why he was so specific about it. It looks like he wanted to do image recognition. The thing with image recognition and deep learning is that, first, you are most likely to use convolutional neural networks. Second, there are many architectures published by big companies, such as Google, and research groups and you might be better of using one of these.

Finally, he wanted the image recognition in order to use it in a recommender system. The truth is that for this particular use case, he could have asked the users to provide tags about images. Hence, image recognition was not an essential part of the problem, which was about suggesting objects to users with similar properties to the ones they liked in the past.

Being too specific when you are not in possession of the full facts can only cause confusion. There are other cases where the same thing can happen. I already gave an example in finance. Another example relates to medical applications.

Let's say that you are interested in sports injuries and you are interested in predicting when a player will get injured. There are different ways to approach the problem:

- Regression—How many days someone will play until they get injured?

- Classification—Break time down in three or four categories (e.g., 1 week, 1 month, etc.) and try to treat this as a classification problem.

- Survival modeling—Similar to regression, but different models and assumptions are used in this case.

The performance of each one of those models will be different, and they might interact in different ways with your problem. For example, classification has the worst resolution (1 week can be a lengthy interval in this case), but easier to solve. Survival modeling is more suited than regression in this case and has more sound assumptions behind it, but there is much fewer survival models than models for regression.

As always, there is no free lunch. Every choice has trade-offs, that's why it is important for a data scientist to be aware of how the results from the models can be used to solve the actual problem.

Pitfalls

Now, we're going to look at some of the pitfalls in the data science process. We'll go over some of the things that can go wrong when a data scientist is trying to solve a problem. We're going to start with what you shouldn't do.

What Not to Do

The first thing you should not try to do is solve a problem by yourself. You might be tempted to try to tell the data scientist what solution you're looking for, but this is not the right approach and you'll see why in a moment.

Your job is to discover what the problem is, phrase it, and explain it to a data scientist following the principles mentioned earlier in the book. Then, you need to take your hands off the problem and delegate.

It might seem trivial to you, but it happens much more often than you'd think. Now, let's take a look at an example to see why this is such an issue.

Example: Bad Collaboration

Let's say that you have an app in the health-tech space and you want to reward users if they take some positive action. However, you also want to see if the app has any real long-term impact on behavior and what that impact is. So, you might ask if someone installs the app, do they keep exercising more or do they eventually go back to a normal level of exercise?

© Stylianos Kampakis 2020
S. Kampakis, *The Decision Maker's Handbook to Data Science*,
https://doi.org/10.1007/978-1-4842-5494-3_9

There are a few different ways to look at this problem. We have a number of variables available to us, such as "before" and "after" activity levels and some other factors, such as demographics. You can either conduct hypothesis testing or you can apply a statistical model. Both options are viable in attempting to tackle this problem.

However, upon closer inspection we realize that the statistical model is a better solution because we have different kinds of variables and it will provide us with a higher level of flexibility than the hypothesis test. Additionally, when the data was collected, there wasn't a control in place which means hypothesis testing might be a bit biased.

So, we run a model and use the installation of the app as a variable.[1] We want to determine whether this variable has any significance or if it has a positive coefficient, which would indicate that the activity levels are higher after the app was installed.

The model might look something like this:

$$ActivityLevels = AppInstallation + DemographicFactor1 + DemographicFactor2 + \ldots + DemographicFactorN$$

We find that the variable *AppInstallation* has a positive coefficient and a significant p-value, which means it has a positive effect.

However, that's when we start running into a variety of issues. The first problem is that the client doesn't understand the model. There are all sorts of terms and concepts they aren't familiar with, and they aren't prepared to learn those concepts.

The thing is that they then need to raise funds based on the research, but they don't really know how to explain what's going on to investors.

At this point, things start to go wrong because the client starts making all sorts of recommendations and effectively tries to solve the problem for the data scientist.

They might start making suggestions for things that have already been taken into account in the model.

For example, they might recommend taking into account other factors as well, but the model already takes into account all the factors that are available from the data.

---

[1] For those of you who want to learn more about the technical details, this problem falls under the "causal inference" umbrella. The actual model presented here is an oversimplified version of the original model, but for the purposes of this example, it gets the message across.

Then, they might suggest measuring the percentage difference against a baseline, but that's something the model is implicitly doing.

Okay, they say. Then, they come up with an even more brilliant idea: let's measure the exact effect our app has on users. Well, that's precisely what the coefficient in the model does.

And finally, they might suggest using a graph because it makes things easier to understand—and graphs always look sexy—even though it doesn't prove anything.

So, we create the graph and it looks like Figure 9-1.

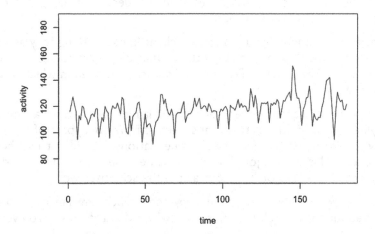

Figure 9-1. Graph of the hypothetical app

The problem is that the graph isn't telling us much.

Statistical models are built to take different factors into account. And in the case of our problem, we have multiple factors interacting with each other. So, when you create a graph, the effect isn't always clear because it's being drowned out by the other factors.

This is why it's a good idea to run a significance test because it can give you a definite answer as to what's happening.

However, the client either doesn't understand the need for a significance test or simply isn't willing to put in the effort. That's when we start going round in circles. The client continues to make similar recommendations and attempts to push the scientist into developing a solution for the problem, but their recommendations aren't methodologically sound.

It's a bad situation to be sure. First, you have a client who refuses to understand the work to present it effectively and, as a result, is forcing the data scientist to produce work that is not scientifically sound.

The most effective solution is to break the problem down into two parts. The first part would involve creating a presentation with some graphs that the client can use to explain the situation to the investors. It might not be completely accurate but it can be useful to raise funds.

Then, the second part would involve creating a methodologically sound model that can be used internally, even if the client doesn't really understand it. That's not necessarily a bad thing. While the client might have trouble understanding the model, it can still be used in the decision-making process to improve certain aspects of the app.

It will make everyone happy because the client has something sexy and shiny to present to investors and to show them what they want to see, but at the same time, they have a scientifically sound statistical model on which business decisions can be based.

So, what was the issue in this situation? The main problem was that the client was trying to replace the data scientist—effectively doing part of his or her job—by coming up with different recommendations. He didn't trust the data scientist to do his or her job, but at the same time didn't understand how statistical testing works or how to evaluate the scientist's work.

These issues were then compounded by the client's unwillingness to learn anything about the model, leading to an awkward situation where everyone lost a lot of time trying to determine the best way to work.

The best approach would have been, of course, to trust in the data scientist without doing this back and forth. The real problem is that this all started because the business didn't have a data strategy in place from day 0.

Had the study been designed properly in the first place, the data would have been collected in a more effective fashion, and that would have ensured the data scientist had the right information to work with. The result would have been one that would have been much easier for everyone to understand, including the investors.

Most data-related problems are due to lack of proper planning. Take this as a lesson and always try to plan in advance as best as you can.

The Real Problem

Data science can be very complex and not everyone will understand it. However, if you try to build something everyone will understand, then you are reducing your work to the lowest common denominator. This is when you end up with something like the anti-vaccine movement.

Imagine what would happen if all the pharmaceutical studies being conducted were run in such a way that everyone could understand them. It simply wouldn't be possible because of the complexity of the information. It would be like handing over the "keys" of a space shuttle to someone who's never even flown as a passenger in a plane. That space shuttle would *not* be going anywhere in the best case scenario, and in the worst, it would blow up in someone's face.

You need people who are experts to validate the study and understand what's going on and to review it. Just as you need a pilot trained in flying space shuttles to get one off the ground.

Clearly, within the context of a business, you don't expect the decision maker, the founder, or the head of a division to be an expert in data science. That's why it's important to find a balance between satisfying the science, so that you know that the model-based decisions are sound, but at the same time making models which are easily interpretable.

As previously mentioned, though, the problems will get worse if a business doesn't follow the right principles from the start, such as not following the right data strategy and so on. Another problem is the lack of data science education, which is why this book is trying to help with.

If someone is properly educated in the process of data science, even if it's only from a high-level perspective, then they will feel more comfortable handling models, interpreting results, and taking action on those results. So, ignorance is a terrible thing, especially when it comes to artificial intelligence and machine learning.

What's the Solution?

While the world of business may not always be data or science-driven, you clearly see the value of data and science-driven decision-making, which is why you're reading this book.

So, how can we solve this problem? Well, first of all, you can't expect every person or organization to be data driven, but there are a few measures you can take to minimize the likelihood of this problem occurring.

Of course, the first and most important thing is to have the right data strategy in place from the very beginning. Something else you can do, though, is to make sure you hire the right people and have faith in them. It's not always going to be possible, but try your best.

Furthermore, don't get overly creative with lots of models and don't try to guide the data scientist into doing specific things instead of letting him or her work.

Data science interacts with business quite often, and I've frequently seen people feel insecure because they don't understand how data science is being used and they try to guide the data scientist into doing all sorts of things. You will waste less money and time if you just hire someone you trust and let them do their job.

You also have to consider the study before you start collecting data. Consider hiring someone even before you start data collection because they can explain how things should be done. So, they can show you how the study should be conducted so that everyone understands what's going on, how to collect the data and that it can be collected in the first place, and that you feel comfortable with the study in terms of understanding it and presenting it to someone else.

As previously stated, in the aforementioned example, most of the problems would have been solved if it would have been easy to carry out a hypothesis test. So, don't be like these people and instead be the person who thinks about all the issues that might come up in advance.

CAUSAL INFERENCE

The model mentioned previously falls under the domain of causal inference. Causality is a very old problem, which has started to become more and more popular in the last few years. Judea Pearl, one of the most important computer scientists in history, creator of Bayesian networks, and Turing award winner (the Turing award is the Nobel Prize of computer science), called for a "causal revolution."

According to Judea Pearl, there are three levels of analysis, which he calls the "rungs of the Ladder of Causation."[2]

In the first rung, we are studying association. We can ask questions such as "if I observed a symptom, how would this change my belief about a disease?"

In the second rung, we are studying intervention. For example, we can ask questions like "if I take aspirin, will my headache be cured?"

In the third rung, we are studying counterfactuals. We can ask questions like "how would my life had been if I had gone to college in a different city?"

Judea Pearl believes that true intelligence lies in the third rung of causation, and one of the reasons that humans have become so successful on our planet is our ability to think in this way. He says that deep learning, while it has been a very successful

---

[2] Judea Pearl and Dana Mackenzie, *The Book of Why: The New Science of Cause and Effect* (New York: Basic Books, 2018).

approach for many problems, is nothing more than curve fitting, essentially being stuck on the first run of causation. According to Pearl, true AI will never be achieved unless we build algorithms that can think up to the third rung of causation much like we do.

Judea Pearl's views have caused criticism and enthusiasm alike. What is certain is that the topic of causality has started to get more and more attention, and it is closely linked to some of the most important questions, such as health outcomes and policy interventions. Microsoft has also released an open source library for causal inference called DoWhy, so we are certain to see more and more research and software libraries coming up in the next few years.

Hiring and Managing Data Scientists

We've talked about why you should hire a data scientist and the numerous benefits it will bring your business. We've also looked at why it's important to let them do their job without you trying to guide them. But to do that, you need to hire someone you can trust.

In this chapter, we'll be looking at precisely how to hire and manage a data scientist to get the best possible results. First, let's try to understand data scientists a little better.

Into the Mind of a Data Scientist

A competent data scientist needs to have a lot of knowledge and a wide range of skills, but certain skills are more important than others.

© Stylianos Kampakis 2020
S. Kampakis, *The Decision Maker's Handbook to Data Science*,
https://doi.org/10.1007/978-1-4842-5494-3_10

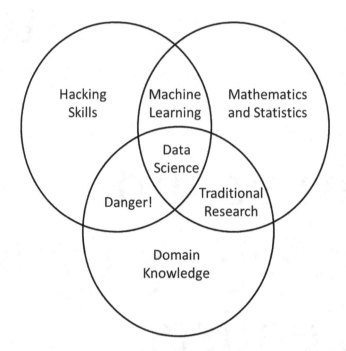

Figure 10-1. The data science Venn diagram

As you can see in Figure 10-1, a data scientist needs hacking skills, math and statistics knowledge, and substantive expertise. While each of the skills is highly valuable individually, if they only have two of the three, they are not a data scientist—and in one particular case, it can actually lead to problems.

Code Hacking Skills

Okay, so when I say code hacking skills, I'm not talking about breaking into NASA, but the fact is that a data scientist needs to have excellent computer knowledge and decent programming skills. This doesn't mean that they need a computer science degree as there are plenty of excellent "hackers" out there who have never taken a single computer course in their life.

However, a data scientist needs to be able to think algorithmically, understand vectorized operations, and work with text files at the command line. Without these skills, it's not possible to be a data scientist. In most cases, the actual quality of the code that data scientists produce does not have to be production level. That's why we are using the term code hacking, instead of software engineering, as the latter means the code should be of very high standard. Data scientists use the code as a tool in order to reach solutions to problems. However, it won't hurt if you find a data scientist whom might also have very good coding skills.

Mathematical and Statistics Knowledge

Once those hacking skills have been used to get the data into shape, it's time to analyze it and extract insights. For this, a data scientist needs to have math and statistics knowledge so they can apply the right methods and models.

Again, a data scientist doesn't need a PhD in statistics to be effective at their job, but they must know how to develop and work with mathematical and statistical models, including how to interpret the results.

Domain Knowledge

Domain knowledge is important for a data scientist: the need to understand the subject matter and not just the technical aspects of it. While this is not a hard requirement, lack of domain knowledge can cause all sorts of issues (we've already seen some examples, and more will follow). Domain expertise allows data scientists to apply the other skills to the data in such a way that they reach the desired goal. Data scientists have to teach themselves the peculiarities of a domain, every time they move to solving problems in a field they've not encountered before.

Two Is Not Enough

Only having two of these skills/knowledge is insufficient. A capable hacker with knowledge of math and statistics but no substantive expertise would be ideally suited for machine learning. However, their skills won't make them a data scientist.

Likewise, if someone has substantive expertise and knows math and statistics, then they are probably looking at traditional research. This intersection is basically what much of academia is made up of—though it should be noted that young researchers are choosing to evolve and learn more about tech. However, without those hacking skills, they will never be a data scientist.

The real problem arises when you have someone who is a skilled hacker and has substantive expertise. It creates a dangerous person because they know just enough to create what seems to be a legitimate analysis but they have no clue how they reached their conclusions or what they've actually created.

Essentially, these are people who have the skills to apply analytics on a problem, but lack the theoretical foundation to make sure they are doing the right thing. So, they might run a statistical test, but could be making erroneous assumptions. However, since they believe they are basing their decisions on data, they have a false sense of confidence which can lead to serious problems.

So, remember, a competent data scientist will be a skilled hacker, have math and statistics knowledge, and also have substantive expertise in your chosen field.

What Motivates a Data Scientist?

Working effectively with someone means understanding who they are and what motivates them. Unfortunately, it can be difficult for people to understand data scientists, precisely because they have such a wide range of skills that they can't be neatly slotted in a box.

A data scientist isn't just someone who knows math and statistics and can work a computer like a pro. These are people who also have great communication and visualization skills, enabling them to communicate well with senior management as well as to tell a story effectively. They are also passionate about the business and curious about data. They constantly want to learn more and they want to solve problems and influence things without the authority that might usually accompany this role. They are strategic, proactive, creative, and innovative people who like to collaborate with others.

They definitely love to be mentally stimulated, and solving problems is their bread and butter. And they have an innate curiosity and drive to grow, which means they love learning new things and acquiring new skills.

The fact is that learning new skills is an integral component of today's market. It's the Red Queen's race or hypothesis in action, which essentially says that an organism must constantly adapt, evolve, and proliferate just to survive.

What Will Disengage a Data Scientist?

If you give your data scientist a lot of repetitive work, you are going to end up with a very disengaged person on your hands. So, if they have to spend ridiculous amounts of time cleaning up data or doing anything else that is the equivalent of counting grains of sand, you are going to be dealing with someone who really doesn't want to be there. And the results will suffer.

Likewise, if they get bored because the problems you are asking them to solve are too simple, you aren't going to win any points. Of course, sometimes simple problems need to be solved, but if you are going to make the leap and hire a data scientist, make sure to take full advantage of their capabilities. Otherwise, it would be like hiring a Michelin star chef to make you a grilled cheese sandwich.

The peril is that the chef, out of boredom, will go off on a tangent and serve you a grilled cheese sandwich that's been deconstructed and reconstructed until it has little in common with that simple sandwich that you wanted.

Also, another problem stems from communication problems. Friction in communication often arises when upper management doesn't understand what the data scientist is doing.

A simple example is that the data scientist ends up doing a lot of data wrangling, which is utterly boring, because of bad communication with the developer in terms of how the data should have been stored.

To return to our Michelin star chef analogy, Heston Blumenthal is a world-renowned chef and his main "thing" is molecular gastronomy. He uses a range of tools and ingredients that are not always the norm in a kitchen. Say you hire him to cater for an event and he tells you what he needs, but because you don't understand his process, you just stock the kitchen with all the usual tools and ingredients.

When he arrives to do the cooking, he ends up having to cook a traditional meal that is completely boring for him, resulting in food that might be decent but nowhere near the quality of what he could have done.

When a Data Scientist Is Looking for a Job

Right now it's a seller's market for good data scientists above junior level where the data scientists are the sellers—you know, seeing as they sell their skills and expertise. The fact is that there is a high demand for skilled people but the supply is nowhere near sufficient to meet it.

For example, an average data scientist in London will receive between one and ten recruiter messages or calls every week. This means they are in a position of power when choosing what projects to work on.

So, to improve your chances of hiring the right person, you need to understand what data scientists want.

What Does a Data Scientist Want?

The first thing is compensation, of course. After all, data scientists are human—even if their skillset and availability can make them seem akin to a mythological beast at times—and they not only need money to live, but they want their value to be appropriately compensated.

However, offering a mind-blowing paycheck alone will not be enough because other things matter too, such as the team they'll be working with, the problem they'll be tackling, the technology stack they'll be using, and the relationship to academia.

The Team

Some data scientists prefer to work in a team where they like the people they are working with. This usually means a geekier, less formal culture or even working from home.

Of course, this won't necessarily work for every company as you can't exactly change your whole culture overnight for the data scientist you are hiring, but you can try to make concessions where possible to improve the working relationship.

The Problem

While we've already discussed the dangers of giving to a data scientist problems that are too simple, you'll also find that many data scientists have a bias for particular types of problems. For example, some might prefer working on problems that have a social impact, while others might stick to problems in fields they like, such as medicine.

Other times, they might have preferences related to the problem itself. For example, some people prefer working with text data.

Often, though, you will find that it is closely connected to doing and learning something new, which is a trait that defines data scientists.

The Technology Stack

The technology being used will also play an important role. Some data scientists will prefer working with stacks they are already familiar with, while others will want to learn new things. However, you will find that most data scientists will avoid old technologies and legacy code like the plague.

Sometimes learning new technologies can be extremely important to a data scientist. It could be because some up-and-coming technologies are dominating the landscape, like Python, and they might want to transition from MATLAB to Python and R, for example.

Or maybe the data scientist wants to expand their skillset because some technologies offer better pay and are more interesting, such as those to do with deep learning.

Relationship to Academia

Many data scientists come out of academia and are used to reading/writing papers and going to conferences. This is an advantage for everyone because conferences can be an essential part of being on top of their game. So, one way to motivate a data scientist is to provide a stipend for conferences.

And, as I already said, it will help you too because your data scientist will learn about the up-and-coming developments in data science, they will learn about fields or applications they weren't previously aware of, and they will catch up with other people in their field.

Some of the data science conferences you can expect a PhD-level data scientist to want to attend include

- ICML[1]—The International Conference on Machine Learning
- NeurIPS[2]—Neural Information Processing Systems
- SIGKDD[3]—ACM SIGKDD International Conference on Knowledge Discovery and Data Mining
- ECML PKDD—European Conference on Machine Learning and Knowledge Discovery in Databases
- AI Stats[4]

Avoiding Traditional Limitations

One frequent problem with hiring data scientists is that employers often take the same approach they would with hiring for any other role in the business. And that usually involves asking if the person has experience in a particular domain.

However, many data scientists, myself included, specifically seek out domains they have not worked in before. It's all about the drive to learn new things and stave off boredom because something you haven't done before will certainly prove more interesting than something you are experienced in.

The important thing that you, as a potential employer, need to understand is that the techniques a data scientist uses have general applicability and the domain itself is of secondary importance. In other words, don't miss out on hiring a great data scientist because they don't have experience in your particular domain.

---

[1] https://icml.cc/
[2] https://nips.cc/
[3] www.kdd.org/conferences
[4] www.aistats.org/

Data Science Is a General Toolbox

The techniques in data science can be applied to a variety of problems. Experience in one thing can translate to experience in something else. So, for example, a data scientist might be able to use the same algorithms to predict disease outbreaks but also how many people might click on an ad.

Think of it like a pastry chef. They learn a wide range of techniques, including working with different kinds of dough, working with sauces, making creams, and so on. And they have a lot of experience making cakes. However, because they are highly experienced with the techniques, they can easily switch from making cakes to making pies.

For example, a common mistake I see made by human resources goes along the lines of

> "We're looking for someone who has 5 years of experience in TensorFlow," says the HR department.

> "Okay, but TensorFlow has only been around for 2 years," replies the candidate or even the recruitment firm.

When it comes to data scientists, their breadth of knowledge is generally a better indicator of their skills rather than just experience.

The problem is that many employers don't understand a data scientist's skillset—not you, of course, because you're reading this book and now have a much greater understanding of data science and what a data scientist does. The result is that many data scientists get stuck in a particular domain because of this lack of understanding.

Let's look at a data scientist who works in finance as an example. In finances, data science would employ tools like time series and predictive modeling. However, these tools aren't limited just to finances. They are equally effective for a number of other fields, such as in sports bettering or even retail. In retail, for example, you can use a time series to predict demand.

However, a large number of employers will see that the data scientist has experience in finance and will automatically assume his or her skillset is only applicable to finances. And this means they could be missing out on an amazing data scientist.

So, don't make the same mistake and give these people a chance, even if they don't conform to the strict model you have in your head of what the perfect hire would be.

Discovering Young Talent

Data science is a little like sports in that talent plays a huge role, especially because it requires a combination of technical skills and soft skills, as we've previously discussed. This person needs to be able to read other people, to make connections others can't see, and to communicate well, beyond the technical aspects of their skillset. They also have to be able to smoothly integrate their soft and technical skills, which isn't always as easy as it sounds.

So, when you find someone with a clear talent for this work, grab onto them with both hands, even if they are at the beginning of their career. Remember, someone who is a junior data scientist today can be your future senior data scientist if you give them a chance and nurture them.

How can you find young data scientists with talent? First, you want to check out the top universities teaching the subject. Then, once you've identified potential candidates, look beyond their university work. You can look on sites like **GitHub**[5] to discover any outside work they've done.

You should also consider creating a mentorship program. This will allow you to identify people with the talent for data science, some of whom you might not have realized have the aptitude for it until you work with them up close and personal.

A Few Typical Data Scientist Dilemmas

When it comes to data scientists, they are faced with a few dilemmas in terms of whom to work for. The options are a startup vs. a bigger company. Each comes with its own set of pros and cons.

First of all, there's the startup. When it comes to a startup, things will certainly be more interesting for a data scientist. They'll be dealing with new technologies and new problems, and they'll have a lot more flexibility. However, the downside is that they'll likely get paid less and stability can be an issue because, after all, many startups don't succeed.

Then you have big companies. The main advantages are that the paycheck will be much more impressive, as will the benefits package. There's also the matter of stability. A big company is unlikely to go belly up overnight and leave a data scientist in the lurch.

---

[5] https://github.com/

On the opposite end of the spectrum, though, the fact is that a data scientist might end up pulling out their hair in frustration because the work will be boring and as far from stimulating as you can get because they'll be dealing with old technology and legacy code, which they'll have to clean up. So, no, it's not going to be fun. At all.

Freeze Your Data Scientist Recruitment Drive Now

If your company isn't ready for data science, you need to stop trying to hire a data scientist. What does it mean to be ready for data science?

Well, first of all, you need to make sure your company has a data-driven culture in place. Everyone in your organization needs to understand the value of data, and they need to be willing to use it effectively.

If everyone thinks data is a waste of time or just something you're doing because it's the latest "fad," then you are going to waste your money and time hiring a data scientist.

Don't worry though, because a little later we're going to talk about how to build a data science culture and you'll know exactly what you need to do.

The next consideration is whether or not you have the data available. If you haven't collected the data, then you might as well wait until you have it. Otherwise, you'll be paying a data scientist to sit around and look pretty. It will also antagonize the data scientist because, as I've already mentioned, boredom is one of the things they hate the most. And a data scientist with nothing to work is a bored data scientist who will likely end up leaving your company faster than you can blink.

Lastly, you need to make sure the data scientist's work will have an impact. Just hiring someone to make data look pretty so you can show off but that's about it is a waste of resources. You need to make sure that you have a clear goal in mind—an important question that needs answering—or you'll be wasting your time and that of the data scientist's too.

When you do hire a data scientist, don't use them just as a highly sophisticated reporting tool. They are a lot more than that and you need to listen to their ideas and trust their recommendations. Don't make the mistake of thinking you know it all.

After all, would you tell your electrician how to rewire your house? Or would you tell your surgeon how to operate on you? Of course not. You'd let them do their jobs and take their advice, because they're the experts. So, why would you do anything differently with your data scientist?

You also need to let them spread their wings. Don't freak out if they want to tackle the problem using a completely new approach. Just because it's not in line with what you had in mind doesn't mean it's not effective. In fact, trying something new is the best way to find the most effective solution.

Remember, doing the same thing over and over again and expecting different results is the definition of insanity. So, you need to give your data scientist the freedom to try something new.

Just put things a little more in perspective for you regarding why you shouldn't be hiring a data scientist if you're not ready; I remember a situation a while ago where a company hired a data scientist. However, the data scientist resigned after only 2 months, and they weren't happy when they did so.

What happened was that the company didn't have any data for the scientist to work on. They were still collecting it, but they jumped the gun, somehow thinking they needed a data scientist on board to look pretty while they were still gathering the information.

Even worse, though, was the fact that they didn't have a clear goal. They just wanted insights, which is way too vague a goal for anyone to do anything with. Of course, with no clear goal, you can imagine that the data collection process was also not up to standards.

To be honest, I'm pretty sure the data scientist didn't quit just out of boredom—though that was an issue. He probably knew he'd have to deal with people who had no understanding of what he did. He'd have to educate them, then force them to set a goal, and then try to turn a bunch of random data into something he could work with. It would have been more hassle than it was worth.

Data Science Tribes

Now we're going to look at the different types of data scientists. I call them tribes because I feel it's a very good word to describe the different groups of people within data science. You don't have just one type of person because you don't have just one way to become a data scientist.

It's not like becoming a doctor where your only path is to go to med school, then do your residency, and get certified. When it comes to data scientists, there are several ways to get to the same end result.

This is why I believe we can basically split data scientists into *three major tribes* and *three smaller tribes*.

In terms of the major tribes, we have the **computer scientist**, the **statistician**, and the **quantitative specialist** who hails from some other field.

The smaller tribes consist of the **self-taught data scientist**, the **software platform user**, and the **domain specialist**.

It's also important to note here that data scientists and data engineers are now considered to be completely different things, even if in the early days the two roles were frequently mixed up and even if data scientists usually have a computer science background but also have some data engineering skills.

The Major Tribes

So, let's take a quick look at the tree major tribes, namely, computer scientists, statisticians, and other quantitative specialists.

Computer Scientists

Computer scientists are the people who have degrees in computer science and then did a master's or PhD in machine learning. The advantages are that they usually have very good skills in coding, databases, and software, but the issue is that they usually ignore traditional statistical techniques in theory, which can be useful in some particular domains and problems.

Computer scientists are very good for issues in machine learning tasks such as predictive modeling. Quite often they'll have experience in Kaggle competitions, which is a prime example of this kind of problem.

Note that Kaggle is a platform that runs predictive modeling and analytics competitions.[6] Companies and users upload datasets and then statisticians, data scientists, and/or other specialists compete to create the best possible models for predicting and describing those datasets. It's a form of crowdsourcing and relies on the idea that countless strategies can be employed in a predictive modeling task and one can't know ahead of time which analyst or approach will be the best. Kaggle is now part of Google Cloud.[7]

You're given a dataset and you just want to find a good algorithm to predict something. The fact that most computer scientists have solid coding skills means it's easier to integrate their work with the rest of the platform, especially if you're a small company or a startup. The majority will be using Python and won't have any problem writing their own APIs. So, essentially, you can use their skills to do many things at once, as they can write the code and integrate it as well.

---

[6]"The Beginner's Guide To Kaggle," *Elite Data Science*, https://elitedatascience.com/beginner-kaggle

[7]Anthony Goldbloom, "Kaggle Joins Google Cloud," *No Free Hunch*, March 8, 2017, http://blog.kaggle.com/2017/03/08/kaggle-joins-google-cloud/

Also, they have solid database skills, so they can do a lot of data engineering work. If you have your own data engineers, it's still worth it because it's easy for these people to retrieve the data, wrangle it into something usable, and so on. This will save your developers time. They might also be able to make suggestions pertaining to the structure of the database, making your life even easier.

However, they have a lack of proper knowledge of statistics so for some particular problems where statistics is appropriate, such as research design, these people can't really help you.

Statisticians

Statisticians are people who usually have a degree in statistics. They might even have a master's or PhD, which can be in statistics or machine learning. They generally have a good theoretical grounding in the field, but they don't have the coding or database skills.

Statisticians have very solid knowledge of statistics and theory. They are the best people if what you are interested in is research design or statistical modeling because you want to know the driving factors behind something.

They're also the best option if you want to model advanced complicated problems while at the same time making sure that the modeling process is transparent, and you understand what's happening.

They are also excellent at taking a critical look at the work of other data scientists because of the rigorous training in math and theory.

The problem is, though, that most of the time they don't really have very good skills in coding and databases. The most likely language they'll be using is R, which they will use as a tool. They probably don't have any experience in other languages or in databases, which will make it more difficult and time-consuming to integrate their work within the system.

Also, some statisticians don't have much training in predictive modeling. On its own, predictive modeling is rarely considered to be important in the context of statistics and is more aligned with machine learning.

So, check if they have any Kaggle experience. It also pays to ensure that they're familiar with some sort of machine learning notions, such as cross-validation, if you're interested in using their skills for predictive modeling.

Other Quantitative Specialists

Finally, we have what I call other quantitative specialists. These are people who come from disciplines heavy in math, such as physics, mathematics, actuarial science, econometrics, and so on. These people might have a master's degree or PhD degree, but this won't be in machine learning or statistics.

These are basically people who figured out that there are more (and better jobs) in data science than in their field of study, and they are trying to change careers. They are trying to capitalize their knowledge in math and/or coding in order to make that happen. You will see a huge variation in terms of skills and knowledge in this tribe.

These quantitative specialists are a bit special because they often have some solid skills and bring a diversity of thinking that's very important in data science. Furthermore, if the problem you're dealing with is in their specific domain, they are definitely amazing people to hire.

However, the problem is that they often lack rigorous training in machine learning or statistics, which can pose a serious problem in certain situations.

So, when it comes to other quantitative specialists, it tends to be subjective in the sense that it depends on the person. However, the main drawback is that these people often lack formal education in the field.

They tend to be self-taught and you need to understand this fact and check to see how effective they are. You can likely find examples of their work on GitHub or Kaggle, for example. And you really want to check beforehand. Please note that being self-taught is not a bad thing—far from it. However, you need to check to ensure their skills are on par with your needs.

These are often people who studied one field and discovered the pay isn't that brilliant or finding a job was virtually impossible. Then they discover being a data scientist comes with a hefty paycheck, so a chemist might decide, for example, to create a few mini projects, upload them to GitHub, or take part in a few competitions on Kaggle in an attempt to find a job as a data scientist.

In my experience, of all the specialties these people have, I've found that the people who are the most effective are physicists. Physics is full of applied math, which makes it easy for them to read machine learning papers, especially since it's, in essence, the same type of math. Furthermore, physicists need to be able to do a little coding themselves, so the skillset is similar to the one required in machine learning.

Something important to note is that domain specialists will have expertise in their respective fields, which might prove useful to you. For example, you might want to hire someone who has some experience in econometrics and a little bit of machine learning knowledge because your problem is related to econometrics in some way. Or, you might want to hire a physicist because you are working with sensor data or radar data.

However, you have to be careful with these other quantitative specialists because some of them will try to fake it. They'll go through a few tutorials and online courses for 2 weeks and, suddenly, they believe they're data scientists. And that's just because they want to get the job. You have to be really cautious of these people because they can derail your whole project and cause a lot of problems.

This doesn't mean that you shouldn't consider people who have little experience but are honest about it and prove they are smart and willing to learn. They want to gain more experience and are intelligent people who have the potential to become outstanding data scientists.

However, in my opinion, it's best to take these people on in a junior position with someone more experienced to oversee them. Of course, the person overseeing them should be a senior data scientist. If you don't have a senior data scientist, hire one first. Then, you can follow their progress as a junior and see how they grow. Depending on their evolution, you can choose to promote them or not.

Convergence Point

It's important to note that after about 5 or 6 years of experience, there's usually a convergence point in all the tribes. In other words, when someone has gained a lot of years of experience and has participated in quite a few projects—let's say between 10 and 30—and they've worked for different companies in various fields, they've usually picked up a wide range of skills. If they have a PhD, that's just an added bonus.

So, they'll have done some coding in both R and Python, and they'll have picked up a little of this, a little of that, and, pretty much, a little of everything. They also have a very good awareness of what's out there. This means that they know all the methods and techniques that are available, even if they don't know how to implement them themselves.

This is important because they know how a problem should be solved. And even if they don't know how to do it themselves, they'll know who to speak to and who to hire to get the work done. Sometimes, it's not important to have the skill per se, but to be aware of one's limitations and have the knowledge of what solutions are available and how to access the people who can implement them.

So, if you see someone who started out as a physicist but has been in data science for 10 years and has worked for a lot of companies in a wide range of fields, then this is the type of person who will likely fall in this category.

Thus, when you find someone who has a lot of experience, you shouldn't worry too much about their background. What's more important is what they've learned over the years and whether they actually have the skills and the awareness of all the techniques and methods that are available.

What you do need to be careful about is when someone has a lot of experience, but they've worked on the same one or two projects for years and years. These are not the type of people I'm referring to. I'm talking about people who have actively tried to learn new things over the years, such as gaining experience with R, with Python, maybe a little research design, and so on and so forth.

The Smaller Tribes

So, now let's take a look at the smaller tribes. When I say smaller tribes, I'm referring to the fact that there aren't as many people in these tribes as there are in the larger tribes. Also, these people tend not to really be data scientists but mainly people with some analytical skills.

Thus, we have the **self-taught** people. These are usually people who studied something random or maybe even software development. They might have participated in a few Kaggle competitions and that's about the limit of their experience.

While the quantitative specialists from the larger tribes might also be self-taught in certain respects, their discipline will have already taught them some skills which carry over to data science.

In this particular situation, I'm referring to someone who studied humanities, for example. An art historian who suddenly discovers their field doesn't offer a lot of professional opportunities or an archaeologist who's discovered that digging through the dirt isn't quite as fun or financially rewarding as they had hoped, and then they suddenly decide to become a data scientist.

Then we have the **software platform user**, which is someone who just knows how to use a specific tool, such as a dashboard, or a software like Weka[8] (which you see in Figure 10-2) or RapidMiner.[9] These people can offer real value for money if you only have a simple problem. For example, if all you need is some reporting, then these people are the best option because they'll definitely be cheaper.

---

[8] www.cs.waikato.ac.nz/ml/weka/
[9] www.rapidminer.com

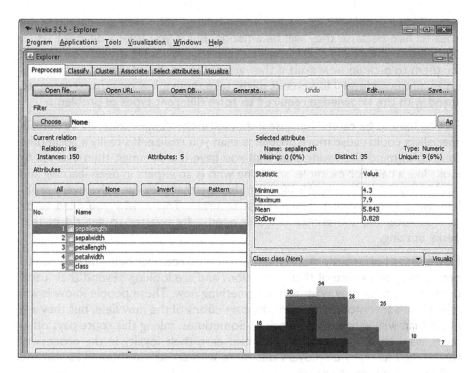

Figure 10-2. The Weka Explorer graphical user interface

They'll also be a lot happier with the work you give them compared to someone with a PhD in deep learning, for example. Otherwise, it would be like hiring a master chef to fry two eggs when as line cook can do just as good a job without being irritated by the work.

However, someone like this won't be able to help you with more complicated problems because they might actually fall into the danger zone we discussed previously and is shown in the Venn diagram at the beginning of this chapter. In other words, they lack a proper background in math and statistics, so they'll be able to apply the tool to certain situations, but they aren't the people you want to trust with building a model or predictive tool.

Last but certainly not least, we have the **domain specialist**. This is someone who has very advanced knowledge of machine learning but limited to a very small niche. One of the most common examples is someone who is specialized in deep learning and computer vision. These are people who have done undergrad work in computer vision or have a master's or PhD in computer vision.

Now, if your problem is computer vision-related, then they are definitely the ones to hire because they will be amazing. But if you have a different issue, these aren't always the best people to turn to. Obviously, it won't be difficult for them to pick up new skills if they are doing machine learning or statistics on an advanced level. However, it might take them a few months to get up to speed with the knowledge required to solve different types of problems.

So, you have to be careful because if they fall in the danger zone, these domain specialists could cause more problems than you realize. It's really a good time to keep in mind the saying that if all you have is a hammer, then everything looks like a nail. For example, someone who is an expert in deep learning and has been doing it their whole life might believe that they can solve every problem using deep learning, which is not accurate. This is why you need to be careful. After all, sometimes you just need a fly swatter to get rid of that pesky buzzing.

However, as I previously mentioned, you should consider being more open to people who are aware of their limitations and are looking beyond their usual domains because they want to try something new. These people know it will take them a few months to learn the particulars of the new field, but they are more than willing to put the work in. Sometimes, taking this route pays off in more ways than one because you'll also earn their loyalty in the process as there aren't as many employers out there who are as enlightened as you and willing to give them a chance.

EXAMPLE: HOW TO EVALUATE A DATA SCIENTIST?

Whom would you rather hire to become your data scientist?

1. Someone who has finished in a top 10 university with an MSc in Machine Learning?

2. A developer who taught himself how to do data science?

3. Someone with a degree in statistics and 10+ years of experience?

The truth is that there is no correct answer. Job performance can depend on multiple factors. A quick search on Google Scholar returns a huge number of results. For example, the search term "IQ and job performance" returns more than 200,000 studies, most of them citing a correlation between IQ and job performance of around 0.5, which is significant. The term "years of experience and job performance" returns more than 4 million results, and the term "job performance predictors" returns more than 1 million results, with papers citing factors from emotional intelligence to personality.

The truth is that it in every case is different. In data science and technology, it is easy to find cases of employees with great credentials, who fail to perform on their role. Cultural fit, personality factors, and personal circumstances can all play a role in an employee's performance.

However, what you can do in order to make sure that employees perform at their best is to create the right environment. This is why in the appendix you are going to find some tools that can help you structure a data science project in a way that you can define clear outcomes and understand what skills someone should have to help you out with this project.

As a rule of thumb, when you are hiring for projects that are non-time critical, then cognitive abilities, IQ, and drive might be the most important factors.[10] If, for example, you have a data science team in place that can help nurture someone with the right drive, then you might be able to get someone on a not very high salary, who will grow alongside the rest of your data science team.

On the other hand, when you are under time constraints, the best approach is to hire someone with the lots of experience and credentials. If the challenge you are facing is very well-defined, then hiring someone who has solved this exact challenge in the past can be a great idea.

---

[10] Richardson, K. and Norgate, S. H. (2015). Does IQ Really Predict Job Performance? Applied Developmental Science, 19(3), 153-169.

Building a Data Science Culture

So, as I promised earlier, we are now going to look at the data science culture. We're going see what a data science culture looks like and what you need to do to start building one in your own company.

An Overview of the Data Science Culture

First, let's take a look at what it means to have a data science culture.

Understanding What a Data Science Culture Is About

Before I get into the actual topic, I'd like to talk a little about the movie *Moneyball*. If you haven't seen this film yet, I urge you to do so because it is relevant to the situation at hand. In other words, it's all about building a data-centric culture within an organization.

© Stylianos Kampakis 2020
S. Kampakis, *The Decision Maker's Handbook to Data Science*,
https://doi.org/10.1007/978-1-4842-5494-3_11

The film is based on the book with the same name[1] and, basically, describes Billy Beane's work at Oakland Athletics. This baseball team was pretty much one of the worst teams in the American baseball championship. However, thanks to Billy Beane's efforts, the team basically goes from the bottom of the pile to being one of the best teams and actually a lead contender.

Beane did this by using analytics and statistics. Essentially, he analyzed players, and even though their individual performance was poor, he saw that together they would make a great team. So, he acquired them, and because of their poor individual performance, other teams thought they were losers and they were cheap to acquire. So, basically, he put together an amazingly good team on a budget.

Throughout the team, you see the resistance Brad Pitt—who plays the role of Billy Beane—encounters because of the traditional ways of thinking practiced by veterans in the club. They simply will not accept the new approach he's trying to implement. Of course, he proves them wrong.

It's a very interesting movie but I won't talk about it further. You should definitely watch it if you are interested in the topic, which I'm assuming you are since you are reading this book.

Now, let's break down the importance of a data science culture.

The Three Levels of a Data Science Culture

A data science culture should be analyzed on three levels, namely, the management level, the employee level, and the organizational level (Figure 11-1).

---

[1] Michael Lewis, *Moneyball: The Art of Winning an Unfair Game* (New York: W.W. Norton & Company, 2004).

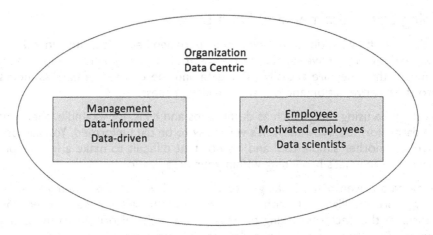

Figure 11-1. A data-centric organization encapsulates data-driven management, a data-friendly workforce, and data scientists

The Management Level

At the management level, it's all about being rational vs. being driven by your gut or taking a traditional approach, like in *Moneyball*. At this level, in a data science culture, the only role guts play is for food digestion, which is what they're there for in the first place.

It's also about being data informed and data driven. We'll look a little closer at these terms in a moment.

The Employee Level

On an employee level, it's all about what we previously discussed, namely, about hiring data scientists and managing them. It's about finding the right people and creating a friendly environment for them. It is also about making sure that the staff that needs to interact with the data scientists (software developers, product managers, etc.) are data aware and believe in the use of data science.

The Organizational Level

On an organizational level, having the right culture means combining the right elements from the management level and the employee level. The result is a **data-centric** organization. This will enable you to maximize the impact of data science from the data collection phase to the research and application phases. That's when you'll start seeing some real results within your organization.

Being Data Informed and Data Driven

Let's start off by looking at the difference between being **data informed** and **data driven**. When we say that an organization or a person is data informed, it means that they are actually using data and the context of data as inputs into their conversation and decision-making process.

So, if you're using things such as dashboards and KPIs, for example, then you are data informed. Nowadays, it's very easy to be data informed. You can use Excel or another simple tool and it's not that difficult to make a few simple charts to see what's happening within your organization.

Being data driven means taking it to the next level. This is when you start using more intelligent algorithms and methods to get the data. Then you transform the decision-making process by using the algorithm to make the decision for you or by taking the algorithm's output into account.

What's the opposite of this approach? Well, the opposite of being data informed or data driven is when you use your guts to make a decision. Or tradition, which isn't much better in terms of accuracy or effectiveness.

Let's take a look at a few more examples.

When you are **data informed**, you might be using a dashboard or Excel to collect data from multiple sources. You've, of course, organized the data appropriately as we've previously discussed in Chapter 2 and have made sure to document the process to avoid any issues. The data is also accessible, and all your employees know about it and are aware of where it can be located and how they can use it.

Being **data driven** means taking things *a step further*. This is where most companies should strive to be in the future. When you get to this point, it means that you are allowing the data and data science to help you make potentially disruptive decisions. In some cases, they might even completely replace human decision-making.

So, if you trade using algorithms, for example, then you are data driven. If your app or web site uses a recommender system, then you are data driven.

Obviously, being data driven does not mean that an algorithm needs to make every decision, but it does mean that you use algorithmic outputs in some parts of the organization and in the decision-making process.

So, why would you want to do this? Being data driven improves efficiency and can actually improve decision-making. For example, a recommender system can make better decisions than humans as to what the client might want. Also, if you want to sell at a massive scale, after a certain point you need to have an algorithm that can actually replace humans when recommending items or curating content. This is just one example but there are many others in different industries.

> ## WARNING!
>
> Eric Peterson calls out a discussion on "What the C-suite should know about analytics" from Kishore Swaminathan, Accenture's chief scientist[2]:
>
> *Data is a double-edged sword. When properly used, it can lead to sound and well-informed decisions. When improperly used, the same data can lead not only to poor decisions but to poor decisions made with high confidence that, in turn, could lead to actions that could be erroneous and expensive.*

In other words, if you use data and data science right, you'll see improved efficiency and better margins. However, if you don't use data in the right way, then it's very easy to end up in the danger zone in Figure 11-2, repeated from the previous chapter.

The problem is that you end up in a position where you use data in the decision-making process, and because you don't have the right theory, or you didn't run the right tests, the decisions are wrong. However, you're still fully confident because you think you used a scientific approach.

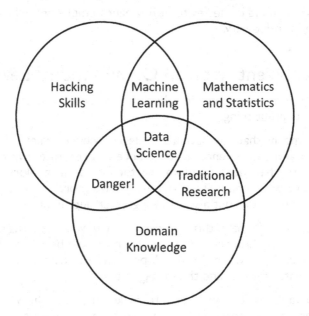

Figure 11-2. The data scientist Venn diagram

---

[2] This interview appeared on Accenture's Outlook journal (2011), issue 1. You can still find this interview here: www.eiseverywhere.com/file_uploads/db8d954d6e3f5fe4a7e67 208537065ac_Accenture_Outlook_What_C_suite_should_know_about_analytics.pdf

The easiest way to mitigate this risk is to hire data scientists who cover the mathematics and statistics circle in the diagram and will use the right theory and processes to do data science. Of course, you need the right culture so that the scientists can interact with other divisions and everyone can scrutinize everyone else's work.

Creating a Friendly Environment for Data Scientists

At the employee level, you want to create a friendly environment for data scientists. We've already covered a lot of this in the previous chapter on hiring data scientists. In essence, you need to remember that data scientists are not only after monetary rewards. They also want the right culture, they want to be mentally stimulated, they want to feel that they are making progress, that they are learning new things, and that their work has an impact.

Also, keep in mind that data scientists tend to be introverts. It's important for everyone to work in teams with people they like and the right culture, but for these people it can be even more important than the average person.

Keeping all this in mind is the key to having happy employees, which translates into increased productivity.

Being Data-Centric on an Organizational Level

An organization is **data-centric** when data science is at the core, which means a few different things.

First of all, it means that you have a 360-degree view of what is happening in terms of data and data science. Data science is a separate division, meaning the data scientist has more freedom and the license to explore new things, even if they are potentially disruptive. They have more responsibility, which can motivate them more but also carry a greater degree of risk.

However, because you have this 360-degree view, you can mitigate this risk. The right communication and reporting channels exist between all important divisions, especially between the developers and the data scientist, but also between the data scientist and the management.

This is important because you want to make sure that the work the data scientist does has an impact and also because you need to be aware of the data that's being collected, how it gets used, and how it's all implemented, which is very easy to get lost in.

To clarify, when we talk data collection, there are two large issues decision makers need to be aware of.

One is legal issues, which have become very important nowadays because of the GDPR in the EU and the United Kingdom. There are all kinds of issues with data privacy, and there's really no excuse not to be aware of how data is being used and how it's collected.

Secondly, it's important to be aware of what the data scientist is researching and also to make sure it's implemented and used in the organization in a positive way. So, this is where having effective communication channels becomes very important. One department can scrutinize the work of another to make sure that what's being done is relevant to the organization.

Let's look at a few examples of how things can go wrong.

Example 1: When Data Is Not at the Core

Let's start with an example of what happens when data science is not at the core. This tends to happen in small or medium enterprises that are trying to reduce costs because they're short on cash.

When a company wants to do data science, but they don't want to hire a new person, they might turn to their own personnel to do data science. They might hire a developer to do machine learning, but that's about it.

It's not a bad idea as someone supervises the software developer. Otherwise, someone might run a model that they got from a library and they don't know how to properly test this model. They might use a cross-validation scheme, for example. The model goes into production and you end up with disastrous results because you're in the danger zone. In other words, you're using a person who has the hacking skills and the domain expertise, but they don't have the theoretical knowledge to back it up.

So, you make decisions that you're very confident in, but they're actually wrong. Technically, it's worse to be wrong and be confident in your mistakes than it is to just be wrong.

Example 2: Things Don't Get Implemented

Another common example that happens mainly in big organizations where there are personnel issues and communication problems is when you hire data scientists and things do not get implemented.

While there are a few ways things might not get implemented, let's look at the case of a data scientist who is not familiar with a domain. So, you hire someone who looks really good on paper. You also really believe in data science, and you give them the freedom and the resources to do whatever they want.

The problem is that it's very easy to create silos in these cases, especially when data scientists speak another language in a sense to other departments. A common mistake is that a data scientist spends months working on something and then it can't be implemented because they were not familiar with the domain. They couldn't understand that what they were creating couldn't be implemented because of real work constraints or due to lack of relevancy from a business perspective.

This tends to happen a lot with people who recently finished their PhDs, and they're still in this very academic mindset of trying to do complicated and fancy things instead of simple things that have an impact.

Thus, having the right culture in data science is not just about being data informed but about transforming creativity into results. It's about giving the data scientist freedom and resources but also making sure the proper checks are in place to ensure the work is relevant and can and will be implemented.

Example 3: Poor Communication

Another example of when things can go wrong is when you have poor communication, which happens quite frequently.

Let's say that there's no one to oversee the communication between the software developer and the data scientist. Even though they are both technical people, they speak a different language most of the time. The result is the creation of silos, which are then difficult to break down. And silos lead to bad practices and increase development times.

First, the problem is the different styles of communication. For example, does the data scientist like to speak a lot using math and different expressions compared to the developers? Also, lack of knowledge of each other's practices is an issue.

Developers like things to be well documented and they like to run tests. Data scientists, on the other hand, can have different approaches. They might even use different programming languages, which can also be an issue.

A solution, in this case, would be to try to make people work in pairs. Another option would be to hire people who have machine language experience, and they can compensate for the poor communication between the data scientist and the developer.

Or you might want to do the opposite and hire a data scientist with some software development experience.

Others have proposed trying to enforce particular software development methodologies across all departments.

The problem is that these don't always work mainly because data science is a bit different to software development and sometimes you can find yourself in an awkward situation. So, the main difference is that progress in data science is not incremental as it usually is in software development. It's a bit more all or nothing.

You might spend a few weeks doing research and then you suddenly get results. Conversely, in software development you might add a new feature every day.

So, even if it's not a bad idea to have stand-ups, for example, where people can share what they're working on every day, don't take into account all the suggestions for software methodologies. You can, of course, think about them, but then adapt them to your specific situation.

MONEYBALL: A STORY ABOUT DATA SCIENCE CULTURE

The world of sports used to be very detached from the world of technology and statistics. Most sports professionals would probably find a discussion about databases or machine learning algorithms pretty boring. Sports was an entirely different world, and Billy Beane's approach seemed alien to the baseball establishment.

Billy Beane had the insight that most of the metrics that were being used in the game were outdated. Players were chosen by scouts who used to be coaches or players themselves. Billy Beane figured out that statistics like the on-base percentage (how often a batter reaches base) were better predictors of success than other statistics that had been traditionally used.

This way, Oakland A's were able to create completely different valuations than the ones that had been used in the past and compete with a budget of around $44 million with teams that were spending nearly three times as much. This approach was called "sabermetrics." The term is derived from the acronym SABR, which stands for Society for American Baseball Research.

Unfortunately, Oakland A's didn't manage to win the championship (even if they got very close to winning it), and then they lost their advantage, once other clubs started copying their approach.

Since then the world of sports has been invaded by data science, with devices and algorithms ruling many aspects of the game. There are conferences in sports analytics being held, with the biggest taking place in MIT's Sloan School of Management (www. sloansportsconference.com).

I have also done research in sports analytics, having worked on the prediction of sports injuries and sports outcomes (like which team will win a game). You can read more on my blog http://thedatascientist.com.

Steps to Build a Data Science Culture

Let's take a look at the steps you need to take to build a culture that is conducive to data science.

Why Do You Need Data Science in the First Place?

First, you need to ask yourself why you need data science. There are three common scenarios in businesses:

1. Data science is an integral part of your business: maybe you're an AI or robotics company, or you're in deep tech in some way, in which case you usually have to have the knowledge yourself.

2. You run a company and want to improve efficiency, so maybe you want to improve operations, for example.

3. You want to improve sales so maybe you want to build your recommender system to improve your B2C offer.

It's unlikely that you fall into the first category or you wouldn't be reading this book. So, you're in either in two or three, and obviously these can be combined.

At this point, you need to ask yourself why you need data science, so you know where you're going next. Now, let's look at resistance to change.

Resistance to Change

If you're a startup, then you're most likely flexible so resistance to change won't be an issue. If you are part of a big organization, though, the following will be very relevant.

The three types of resistance to change are cultural, personal, and intellectual (Figure 11-3).

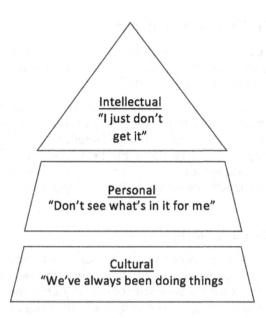

Figure 11-3. Types of resistance to change

Cultural Resistance to Change

Cultural resistance to change is what we see in *Moneyball*. We have a big and old organization or multiple layers of management or fields that haven't had much to do with math, statistics, or data. A good example is the comparison between sports and finance.

As shown in *Moneyball*, the world of sports is still trying to get deeper into analytics, but it's just not there yet. Even back when the events of *Moneyball* took place, the kind of things that Billy Beane tried to implement in Oakland Athletics are nothing special in finance, which is usually one of the fields at the forefront of data analysis.

Personal Resistance to Change

Personal resistance to change also occurs in smaller companies and is when employees don't like the change.

To give you an example from personal experience, I worked with a client and waited 2 months for someone to give me a database expert. Then, I had to call the supervisor and wait another month.

What may have happened is that the employees could have been over-worked and they were handed even more work. So, the employees don't really see any benefit out of doing anything, like exporting a database or trying to implement an algorithm.

Intellectual Resistance to Change

Finally, we have the case where people are afraid of what they don't understand. Think Skynet in *The Terminator* movie, as in when people are terrified that machines will take over the world.

There isn't a lot of intellectual resistance to change in organizations because most understand that data can be useful since it can improve efficiency and sales. However, if there's the slightest hint of intellectual resistance to change, this is where hiring a good data officer can be a good start. This is someone who can own the data and can show it to people, as well as showing how it can improve pretty much everything within a company.

The Journey to Change

When it comes to your journey to change, there are three important points to remember. First, you have to determine how data scientists will get you there. This means creating a data science–friendly environment where they can talk about it and trying to run projects with a large potential impact. Change is slow, but if people start seeing results, then they'll slowly change their minds because they will understand the benefits.

It's very important to also find a champion in some cases. This is someone who can create enthusiasm in this new direction. The champion is someone who truly believes in this new approach, and they have the enthusiasm and the status to actually push the rest of the company toward this new goal.

It's usually a high-ranking employee who is also well liked. It could be you or it could be a data scientist or the chief data officer.

Generally, the best option is for it to be someone who is already in the organization and is popular, someone people respect and listen to regardless of what they're talking about. This could also help employees see this as a learning experience and something more positive for the company and hence for themselves.

The final point to remember on your journey to change is that it's important to focus on benefits for the employees because everyone hates having to do more work with no reward. However, employees who are intellectually curious and driven will see the benefits of learning and also helping the organization since this can translate into results for themselves.

How to Start

When it comes to implementing a data science culture, I've seen two approaches. The first is the baby steps approach, and the second is the use case approach.

In the baby steps approach, you collect lots of data and the decision makers educate themselves in data science. Then, they try to use simple graphs and stats, so they become more data informed. Once they feel comfortable with the data, they take the next step and hire someone to make sense of the data.

The use case approach is more of a top-down approach. This is when a company has a certain type of problem, and they set up a plan and might start by hiring someone to help them design the plan and solve the particular problem. Then, they'll consider how the findings are relevant to other use cases.

The baby steps approach is the most common because it's easy to collect and visualize data in Excel or maybe a dashboard. This happens when a company is interested in data because they hear that everyone else is collecting and using data, but it's not really a priority.

So, at one point the company has solved its main problems and is profitable, and they decide to hire someone because they have all this data and want to do something with it. This approach makes sense because it's not too disruptive to current operations, and if data collection is not your priority, you don't have to push things too far, too fast.

However, it can also cause all sorts of problems. It's a slower way of adopting data science, but then you also get all the issues we talked about in Chapter 2. In other words, a bunch of data is collected with no prior thought or goal, and it turns into a disaster.

The use case approach is less common, and it happens when the company has a clear use case. The advantage is that it can provide more immediate results using a more targeted approach. However, it requires some level of data awareness and the company needs to have access to a data scientist they can trust.

Note that a company doesn't have to choose one approach over another. Quite often they might start with the first approach and end up with the latter.

Where to Start

The where is just as important as the *how*. So, you first want to map your organization's use of data. You need to understand the flow of data from collection to usage and discover any dark data you might have.

Also, emphasize strategy and innovation and try to help employees understand why this is important. Reward good behavior, which is very important for data scientists. Many data scientists want recognition and to feel as if they are making an impact on the world or in their company. Thus, promoting good work on a blog and social media will help a lot, as well as sharing the best use cases with everyone in the company.

Finally, creating an embedded culture is the way to break down silos and improve communication. You want data scientists to work closely with developers, the product owners, and the managers.

This is the only way to make sure that the research is relevant, that the results are implemented, and that they have an impact within the company.

Understanding and Using Dark Data

Dark data is operational data that is not being used, such as log files, notes from sales calls or emails, and similar information.

This type of data is relevant because there's a lot of it and it's sitting there gathering dust. Deloitte reports that by 2020 we're going to have 44ZB of data.[3] This figure might not be 100% accurate, but it's still a huge amount of data. Note that many of the sources of data aren't specified; it includes audio, notes, video, image files, and so on.

Also, Deloitte has a special mention for the Internet of Things and other non-traditional sources. This type of data is easy to collect but it's not always used. So, it very quickly becomes dark data and just sits in a data warehouse waiting for someone to make sense of it.

What Can Dark Data Do for You?

There are a number of things dark data can do for you, but it's not easy to provide a general set of guidelines that will apply in every situation. However, here are a few examples.

So, you might have some email address. In some cases, you can extract gender and age from email addresses or visitor behavior from log files or sentiment from customer calls. Then, you can create predictive models for user retention from the usage logs or customer logs, for example. This is just one example and there are countless possibilities.

Keep in mind that the unstructured nature of data can make it difficult to use. However, if you can do it, it can be very rewarding.

---

[3] www2.deloitte.com/content/dam/Deloitte/is/Documents/technology/ deloitte-uk-tech-trends-2017-dark-analytics.pdf

Other Steps to Take to Become More Data Driven

As mentioned, you need to create an open culture that promotes strategy and innovation. A champion can be a vital asset as they help motivate employees.

There are a number of ways you can create this kind of culture. You can, for example, organize internal hackathons once or twice a month or allow employees to play around with an idea of their own. Other companies have done this successfully, and it works because it keeps people happy and engaged.

Another interesting approach is to use Kaggle to create internal competitions. This works if you have a very large number of data scientists within your company. It gives people an opportunity to have some fun and learn something new.

Finally, if you can, open up your data and allow your employees to invite other people to take part in small projects. Of course, this doesn't apply for every company. However, if you can, it can be a great way to find new talent.

Rewarding Good Behavior

When it comes to rewarding good behavior, employees with good ideas need some form of recognition. Maybe you publish their results on a blog, or you make libraries and software open source and attribute the developers. This is what large organizations in the field are doing.

For example, Google[4] and Facebook[5] have their own research blogs. Facebook also has a forecasting library called Prophet[6] when you can read an article written by the people within the company who created the library.

This is an excellent approach because everyone wants to be rewarded for their contribution to something new and amazing. And there's no reason for them to be anonymous. Their names are right there. Plus, data scientists like to look into the future. If they want to get another job, it's good for them to know that a new employer can easily find them and their work online. This is why data scientists and software developers also have GitHub accounts.

So, it's not just about the bonuses, it can also be about the ego.

---

[4]https://research.googleblog.com/
[5]https://research.fb.com/blog/
[6]https://research.fb.com/prophet-forecasting-at-scale/

Create an Embedded Culture

Finally, one of the most important things to do when trying to build the right culture for data science is to create an embedded culture. This means that you want the data scientist to work closely with other divisions and you want to have good communication between all parties, as well as full transparency.

Stand-ups where people share what they are working on are a pretty standard way of promoting communication and transparency. However, be careful because daily stand-ups can get irritating because of the slow nature of a data scientist's work. It isn't incremental and the last thing you want is to have a bunch of introverts feeling uncomfortable because they have no results to show. This is why weekly stand-ups might be a better option.

You also should consider mentorship programs for new data scientists. Junior data scientists can be very enthusiastic, and sometimes you can find amazing talent from some good universities who might be a bit inexperienced. This might be their first real job, so it's a good idea to have a senior data scientist educate them and mentor them, which will allow them to become great.

CASE STUDY: BOOKING.COM DOES DATA SCIENCE THE RIGHT WAY

Booking.com is a company that does data science right. They sell over 1.5 million hotel rooms every 24 hours, and millions of people browse the site constantly, which means they have massive amounts of data. They are continuously using that data to improve their services and develop new ones.

Booking.com has over 120 data scientists with different profiles, backgrounds, and working preferences. The culture is clearly data driven and has been developed in such a way that the company offers the perfect venue for data scientists. From challenging them with different projects that require them to constantly learn new things to helping with their education through workshops and presentations held by their peers, there is little opportunity for a data scientist to get bored at Booking.com.[7]

The structure is also one you want to emulate as it's embedded, like we previously discussed. Their data scientists are an integral part of the business. They are part of teams that include developers, product owners, and various other specialists. The team is formed in such a way that they can take an idea from concept to the implementation stage.

---

[7] Nishikant Dhanuka, "Diary of a Data Scientist at Booking.com," *Towards Data Science*, November 13, 2017, https://towardsdatascience.com/diary-of-a-data-scientist-at-booking-com-924734c71417

On a daily basis, they follow what they call the booking agile essentials recipe, which means they have daily stand-ups, backlog, retrospectives, KPIs and objectives and key results, team purpose, and more. This approach allows their teams to move forward or to fail and learn from their failure as quickly as possible.

For this reason, data scientists at Booking.com are expected to have excellent communication skills as well as commercial awareness. In other words, technical skills aren't enough.

However, data scientists are happy at Booking.com because they are constantly working on new projects and learning new things. They don't have to wait around for data to show up, their advice and recommendations are taken into account, they are allowed to take new approaches, and they are treated like the data scientists they are.

So, you could do worse than to emulate the Booking.com model in terms of creating a data-centric culture. Of course, you don't have to do everything they do. It is a huge company after all. Maybe, in your case, you don't care so much about communication, but are more interested in the technical aspect.

However, if you want to create an embedded culture, then definitely look into how Booking.com is doing it.

Data Science Rules the World

Data is being created and consumed at an unprecedented scale. Data science is progressing even faster, further speeding up the rate of data creation and consumption.

It is clear that data science is expanding both vertically and horizontally across all industries and layers of society. The companies that are the first to adopt the best practices will gain a significant advantage. The rest might just perish. The situation is to a large extent equivalent to a paradigm shift in the means of production, like the one that was experienced through the industrial revolution, making many people talk about a fourth industrial revolution.

The history of data science is not linear. Its roots are in mathematics and philosophy, in the vision of the AI fathers to create a thinking machine, in computational intelligence and data mining, and in many other fields and ideas as to how we can extrapolate from a simple piece of information to describe a universe of concepts. After centuries, the science, engineering, and availability in data and computing power have created a perfect storm, and this storm is only going to get stronger.

I hope this book provided you with enough insights as to how data science can be applied in business. Data management, data strategy, choosing the right algorithms, hiring data scientists, and having the right culture all these factors are interleaved when a company attempts to extract insights from

© Stylianos Kampakis 2020
S. Kampakis, *The Decision Maker's Handbook to Data Science*,
https://doi.org/10.1007/978-1-4842-5494-3_12

data. Doing data science is more complicated than simply hiring a data scientist, but, at the same time, it can be much easier than creating a novel algorithm from scratch.

Being aware of these facts is the first step toward building a data-centric organization. This book has provided you with enough material, examples, and case studies to be aware of the opportunities, but also the pitfalls that may arise. Maybe you've not absorbed all the lessons now, but that's fine. Not everything might resonate at first instance. After you start working with data, you will revisit some of the chapters, and then they will make more sense, as you will be able to attach your experiences to the teachings inside this book.

What's important is that you've decided to stay and be part of this revolution by learning how to use data science in your company.

I wish you success on this journey.

Tools for Data Science

The Data Strategy Canvas

The data strategy canvas is a tool that is used in order to structure the implementation of a data science within a company. You will find this canvas especially useful if you are implementing data science for the first time within your company.

The canvas is covering the most important topics that you have read in this book:

1) Challenge—Here you have to specify the exact problem you are trying to solve. Don't forget to pose the problem as a question. This forces you to make the problem more concrete and will also help the data scientist better understand what it is that you are trying to do.

2) Data sources—Where are you getting your data from?

3) Appropriateness, nature, time, and cost—These are the four data considerations that were discussed in Chapter 2 about data management.

© Stylianos Kampakis 2020
S. Kampakis, *The Decision Maker's Handbook to Data Science*,
https://doi.org/10.1007/978-1-4842-5494-3

4) Method—While a data scientist would be able to better advise you on this, it is still a useful exercise to try to think how you could solve this problem. It will at least help you make potentially better hiring decisions. However, do not get too stuck on your answer to this question. For example, it is likely that something that might look like a deep learning problem to you can be solved more easily with different techniques. Discuss with the data scientist the best approach, and let them have the final say.

5) Success criteria—When will you know that the project has finished? The wrong expectations can cause all sorts of problems. If you can't define clear success criteria, then at least think whether you can break down a project into smaller milestones that are more manageable.

6) People—Use the things you learned from the "Hiring and Managing Data Scientists" section in order to figure out which type of data scientist would be better suited to this problem. Also, think whether it is best to work with a contractor or a full-time employee. Many projects are one-off, so contractors can be a good choice.

7) Culture—Will a data scientist feel comfortable working with you, or will they feel like second-class citizens? You might spend a long time finding the right person, and then they might leave if the company does not provide the right cultural fit.

The data strategy canvas

Challenge
Data sources
Appropriateness (can the data be used to solve the problem?)
Nature (is the data noisy or other issues?)
Time (how fast can you acquire new data?)
Cost (how expensive is the data?)
Method (how will you solve the problem, machine learning, deep learning, statistical modeling, etc.?)
Success criteria
People (whom should you hire?)
Culture (do you have the right culture for data scientists to come and work for you?)

The Data Science Project Assessment Questionnaire

The purpose of this questionnaire is to help you clarify your objectives and factors that can affect the success of a data science project.

It is important to understand that a data science project is often an exercise in risk management. The problematic part in data science is the "science" bit. Quite often, it is impossible to know how well something will work in advance, until you try things out. The only cases where you can be confident of results before you embark on a project are

1) You have worked on the same case with very similar data in the past.

2) There is an extensive body of literature on this type of problem.

In any other case, the results cannot be guaranteed. Hence, the right project structure should focus on organizing an "attack plan," where each step should produce one of either two outcomes:

1) Successful completion of the project.

2) Learn something about the problem (e.g., maybe a family of methods is not appropriate, which can lead to improved exploration)

In order to better manage the project, expectations, and risk, the following questionnaire can help you list all the important requirements. This questionnaire is both for you and the data scientist, and it will help you build a common understanding of the requirements, the challenges, and how to mitigate risks. It is recommended that you do at least one round of back-and-forth between you and the data scientist in order to make sure that there is appropriate understanding on both sides.

The Data Science Project Assessment Questionnaire

Success criteria

What would you consider successful implementation of the project? Share your own thoughts.

Are there any benchmarks in performance? Please try providing a numerical answer (e.g., "anything above 60% accuracy is good, based on a benchmark X").

Risk factors

Is the data of appropriate quality? If no, what are the issues?

Is the data of appropriate size? If no, how much do you think this will affect performance? How will you mitigate that risk?

If the data is of high quality and size, then what could possibly prevent a good model from being built (e.g., maybe the domain is particularly difficult)?

Timelines

How time critical is the project?

If the project is difficult and no approach can reach the desired accuracy, which of the following plans seem the most attractive to you and why?

- Keep on trying more advanced approaches (e.g., ensemble modeling).
- Fix data issues and try again (e.g., collect more data).
- Simply use a model that is "good enough," even if it doesn't reach the desired performance.

Best/worst possible case

What would be the best possible outcome for you?

If the goals cannot be achieved (e.g., the model doesn't work to the desired performance or fails completely), what would you do?

Interview Questions for Data Scientists

Interviewing data scientists is not easy for many reasons. As the book explained in Chapter 10, there is not a single data science curriculum that someone should abide by. You get tribes of data scientists that can have different mentalities and use different tools. This makes interviewing data scientists very challenging. This becomes even more challenging when someone is your first hire.

These questions are designed to help you understand how good someone is.

What are the different types of machine learning? Can you explain them to me?

This is a very simple question. Someone who has spent time reading the subject will know that the main types are supervised learning, unsupervised learning, active learning, semi-supervised learning, and reinforcement learning.

Active learning and semi-supervised learning are not very popular, so maybe not everyone knows about them, but you can give bonus points to someone who does.

What is the difference between descriptive and inferential statistics?

Descriptive statistics refers to the statistics that most people do in high school, like summary metrics and graphs. Inferential statistics deals with the problem of inferring something about a population by only getting a sample out of it. Statistical modeling and hypothesis testing, they are all part of inferential statistics.

Which algorithm would be better in a given problem: Random forest or Naïve Bayes?

This is a trick question. Random forest is a much more successful and popular algorithm than Naïve Bayes. However, the no-free lunch theorem tells us that there is the possibility of simple algorithms being better than more complex ones in a given problem. This relates to the inductive bias that an algorithm might have.

While a full technical exposition of this argument is beyond the scope of this book, you will essentially expect two kinds of answers. Data scientists that lack the proper theoretical background will say "random forest." When asked why, they might not be able to properly justify their choice or say something like "it's a good algorithm." Data scientists that fully understand the theory behind machine learning might say something like "it depends" and will give a more thorough answer, which will demonstrate that they understand that a complex algorithm is not always the right choice.

You have a dataset of 1000 rows and 1500 variables. Which algorithms would be suited to this problem?

This is a particular type of challenge where the variables are more than the rows. These problems can be solved through algorithms that are good in handling a large number of variables. Some choices include

1) Elastic net

2) Random forests

3) Gradient boosted trees

4) Heavily regularized neural networks

If someone misses this question, it is not very important, but answering it probably demonstrates solid understanding of some important concepts.

Is more data always good?

This is a trick question. While more data in terms of rows is good, more variables is not always a good thing. There is a concept called "the curse of dimensionality." According to this, adding more variables can make sometimes a problem more difficult to solve, especially when the variables do not contain much information. Hence, you expect that someone who has the right background to answer that more rows of data are good, but more variables might not be necessarily good and whether they are good depends on the problem and the quality of the data.

Other things to ask about

GitHub repository, or samples of code. Samples of code are actually better indicators of someone's skill and interest in the area than abstract coding exercises, like the ones you see in most interviews.

Kaggle. While many brilliant data scientists might not really have the interest or time to participate in Kaggle competitions, having a Kaggle account is always a good thing.

A degree from a top university.

Side projects that might relate to your particular challenges.

The New Solution Adoption Questionnaire

It is often easy to get carried away by new technologies. Every 1–2 years, there is some new big name in town, and all the companies are racing to adopt the new tools. However, quite often, those solutions are expensive to implement, might not carry any benefits to you, and might be quite immature and untested. There have been countless businesses that rushed to use Hadoop, NoSQL, blockchain, and other technologies without really needing them. This simple questionnaire should help you understand whether you should adopt a new solution or not.

Step 1

Goal: Understand your objective and what you are trying to achieve. For example, what if you want to use a new kind of database, what are the issues with the current one? Long read times, scaling up, or something else?

Step 2

Enumerate each solution, context it was created in, and pros and cons. Read the white paper if there is one available. For example, I've filled in the responses for the Ethereum blockchain in the top row of the following table.

Solution	Goal of the solution	Creator/maintainer	Pros	Cons
Ethereum blockchain	Decentralization, immutability, smart contracts	Ethereum Foundation	Very well tested solution	Blockchain solutions face speed issues (whether this is relevant depends on your particular challenge)

Step 3

Analyze costs and risks. Every migration or adoption of a new technology contains unseen risks. Answer the following questions to map out risks and solutions:

Can your current engineers do this or you need to hire more people?

If you need to hire someone new, how easy is it to find someone, and how much would it cost?

What could go wrong? How can you mitigate the risks?

How much money would you save after the solution would be implemented?

How would your service or product improve by adopting the new technology?

Index

A

Anomaly detection, 85

Artificial intelligence (AI), 4
  automated planning, 7
  general, 21
  history, 5
  research, 10, 11
  vision, 6
  winters, 9, 10

Automated planning, 7–8

B

Bayesian vs. frequentist statistics, 74, 75

Big five personality traits, 86

Booking.com case study, 140, 141

C

Causal inference, 102–103

Cognitive science, 5, 6, 11

Computational intelligence (CI), 17, 18

Computer scientists, 116, 117

D

Dark data, 138

Data acquisition
  choosing data, 25
  cost, 26
  nature of data, 25
  problems, 26, 27
  time requirement, 26

Data-Centric, organizational level
  data collection, 130
  data, not at core, 131, 132
  data privacy, 131
  poor communication, 132

Data collection
  B2C apps, 32, 33
  finance, 34
  retail, 33
  sales, 33
  social media, 34
  sports, 34
  types, 24

Data management, 34
  bad practices
    case study, 38, 39
    connection, 36
    documentation/data standard, 36
    lack of objective, 36, 37
    objective, 36–38
  definitions, 23
  good practices, 39
    awareness, 35
    data standard, 35
    establish, goal, 35
    issue, 40
    setting goals, 41
  sources of data, 24

© Stylianos Kampakis 2020
S. Kampakis, *The Decision Maker's Handbook to Data Science*,
https://doi.org/10.1007/978-1-4842-5494-3

Printed in the United States
By Bookmasters